Electrical and Electronic Principles
Second Level

Hutchinson BTECtexts

Learning by Objectives
A Teachers' Guide
A. D. Carroll, J. E. Duggan and R. Etchells

Engineering Drawing and Communication
First Level
P. Collier and R. Wilson

Physical Science
First Level
A. D. Carroll, J. E. Duggan and R. Etchells

Workshop Processes and Materials
First Level
P. Collier and B. Parkinson

Electrical and Electronic Principles
Second Level
W. Bolton

Electronics
Second Level
G. Billups and M. T. Sampson

Engineering Drawing
Second Level
P. Collier and R. Wilson

Engineering Science
Second Level
D. Tipler, A. D. Carroll and R. Etchells

Mathematics
Second Level
G. W. Allan and A. Hill

Site Surveying and Levelling
Second Level
J. Pettet

Digital Techniques
Second/Third Level
C. Kelly

Control of Manufacture
Third Level
W. Bolton

Electronics
Third Level
G. Billups and M. T. Sampson

Communication Skills
P. Panton

General Studies for Technicians
P. Denham, H. Bamforth and J. Derbyshire

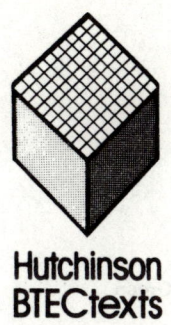

Hutchinson BTECtexts

Electrical and Electronic Principles

Second Level

W. Bolton

Hutchinson

London Melbourne Sydney Auckland Johannesburg

Hutchinson & Co. (Publishers) Ltd
An imprint of the Hutchinson Publishing Group
17–21 Conway Street, London W1P 6JD

Hutchinson Publishing Group (Australia) Pty Ltd
16–22 Church Street, Hawthorn, Melbourne, Victoria 3122

Hutchinson Group (NZ) Ltd
32–34 View Road, PO Box 40-086, Glenfield, Auckland 10

Hutchinson Group (SA) (Pty) Ltd
PO Box 337, Bergvlei 2012, South Africa

First published 1985

Set in Times by Activity Ltd, Salisbury, Wilts

Printed and bound in Great Britain by
Anchor Brendon Ltd, Tiptree, Essex

British Library Cataloguing in Publication Data
Bolton, W.
  Electrical and electronic principles,
  second level.
  1. Electric engineering
  I. Title
  621.3   TK145

ISBN 0 09 159411 1

Contents

Introduction   8

1   Basic circuit theory   **9**

Revision of basic principles   9
Multiples of units   16
Series circuits   19
A potential divider   21
Parallel circuits   22
Series-parallel circuits   26
Electromotive-force (e.m.f.)   30
Kirchhoff's laws   31
Power in a d.c. circuit   36

2   Electric fields and capacitance   **40**

Charge   40
Electric fields   43
Electric potential   44
Capacitance   48
Capacitance of a parallel plate capacitor   51
Multiplate and variable capacitors   54
Dielectric strength   55
Energy stored in a capacitor   57
Capacitors in parallel   59
Capacitors in series   59
Series-parallel connections   62

3   Basic electromagnetism   **65**

Revision of basic principles   65
Magnetic forces   67
Coils in magnetic fields   70
Magnetic flux   74
Electromagnetic induction   75
The induced e.m.f.   80
The simple a.c. generator   82

4   Magnetic circuits and materials   **85**

Magnetic circuits   85

Series connected magnetic circuits   91
Flux leakage and fringing   95
Magnetic field intensity   96
Magnetic materials   98
Magnetization curves for ferromagnetic materials   99
Hysteresis   104
Magnetic screening   107

5   Inductance   **108**

Self inductance   108
The inductance of a coil   111
Arcing in inductive circuits   113
Energy stored in an inductor   114
Mutual inductance   115
The transformer   116

6   Alternating voltages and currents   **120**

Waveforms   120
The sine wave   122
Average values   126
Root mean square values   129
Form factor   132
Phasors   133
Addition of phasors   135
The sinusoidal waveform equation   138
Rectification   139

7   Single phase a.c. circuits   **141**

A pure resistor in an a.c. circuit   141
A pure inductor in an a.c. circuit   145
A pure capacitor in an a.c. circuit   148
Inductance and resistance in series   150
Capacitor and resistor in series   156
Power and power factor   160
The power triangle   164
The series LCR circuit   166

8   Semiconductor diodes and transistors   **172**

Resistivity   172
Conduction in solids   174
Atoms   175
Semiconductors   176
The junction diode   178
The bipolar transistor   180
Transistor circuit configurations   183

9   Measuring instruments and measurements   **186**

Types of measuring instrument   186
The moving coil instrument   187
Changing the range of a moving coil instrument   189
The moving coil instrument as an ohmmeter   192
Multirange moving coil instruments   193
Moving iron instrument   195
Voltmeter sensitivity   197
Using voltmeters and ammeters   198
Instrument errors   199
Dynamometer instruments   200
The cathode ray oscilloscope   201
Electronic voltmeters   204
Digital voltmeters   205
The potentiometer   205
The Wheatstone bridge   208

Solutions to self-assessment questions   210

Index   229

Introduction

This book is intended for all those technicians who want a basic grounding in electrical and electronic principles to complement an electronics or electrical engineering technology. It covers the Business and Technician Education Council unit U81/747 Electrical and Electronic Principles II. This unit is an essential study in all Certificate and Diploma syllabuses for Electrical and Electronic Engineering programmes, and many others. The book essentially follows the sequence of that unit.

Each chapter is headed by the aims for which the chapter was written. Within each chapter, each significant segment of material is accompanied by self-assessment questions. The solutions for all these questions appear at the ends of each chapter. In addition a large number of worked examples have been included.

Because of the way the subject matter has been carefully developed and the copious use of self-assessment questions, the book could be used as the basis of an open-learning system.

* Basic electrical and electronic principles
* Covers the BTEC unit Electrical and Electronic Principles II
* Includes over 140 worked examples
* Includes over 280 self-assessment questions
* Solutions given to all questions

1 Basic circuit theory

After reading this chapter you should be able to:

1 Use the terms potential difference, current and resistance, their units and multiples of them.
2 Draw and interpret simple circuit diagrams involving series and parallel arrangement of resistors and sources of e.m.f.
3 Apply Ohm's law to circuits involving series and parallel combinations of resistors.
4 Explain how series resistors can be used as potential dividers.
5 Distinguish between the e.m.f. of a cell and the potential difference between its terminals, using the term internal resistance.
6 Apply Kirchhoff's laws to circuit problems involving not more than two unknowns.
7 Calculate the power dissipated by resistors and circuits.

Revision of basic principles

This chapter is about currents in electrical circuits. The circuits are considered to only contain resistance and a d.c. source or sources. The following brief notes are intended as revision of the basic terms and principles associated with such circuits.

For an electrical current to flow between two points there has to be a potential difference between those points. If there is no potential difference there is no current. We can imagine the situation to be similar to a trough containing water (see Figure 1). The water will flow along the trough if there is a difference in height between the ends of the trough. If there is no height difference there is no flow. The flow of water represents the current, the difference in height the potential difference. If there is a potential difference between any two points in an electrical circuit and a path for a current exists between those points, then a current will flow.

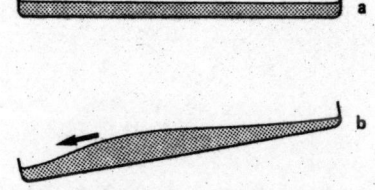

Figure 1 (a) *Ends of the trough at same height, no movement of the water*
(b) *A height difference produces a flow of water along the trough*

In an electrical circuit, a current (symbol I) can be detected and measured by an instrument called an ammeter. The unit of current is the ampere (A). Potential difference (symbol V) can be detected and measured by an instrument called a voltmeter. The unit of potential difference is the volt (V). The term voltage is sometimes used in place of the term potential difference.

For a particular potential difference between two points in a circuit the current, for a d.c. source, will depend on the resistance between those points. Resistance (symbol R) is defined as:

$$\text{Resistance } R = \frac{\text{Potential difference } V}{\text{Current } I}$$

$$\boxed{R = \frac{V}{I}}$$

The unit of resistance is the ohm (Ω).

The term resistor is used for a circuit component which has been deliberately introduced into an electrical circuit because of its resistance. Such a component may be just a coil of wire or a small block of conducting material.

For many resistors the relationship between the current and the potential difference is a straight line graph. The line passes through the point where both current and potential difference are equal to zero (see Figure 2). The slope of such a graph is constant, i.e. V/I is a constant. This means that such a resistor has a resistance which does not change when the current changes. So, if we know the potential difference at one current, we can calculate the potential difference at another current.

For example, if the potential difference was 3 V when the current was 1 A, then the potential difference when the current is 2 A can be calculated. Initially:

$$R = \frac{V}{I} = \frac{3}{1}$$

At the new current:

$$R = \frac{V}{I} = \frac{V}{2} = \frac{3}{1}$$

$$\therefore \quad V = 6 \text{ V}$$

In this example the current was doubled, from 1 A to 2 A. The result was that the potential difference doubled, from 3 V to 6 V. This is always true for this type of relationship. The potential difference is proportional to the current:

$$V \propto I$$

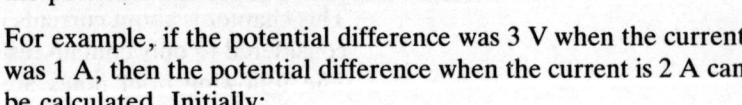

Figure 2 *A potential difference–current relationship*

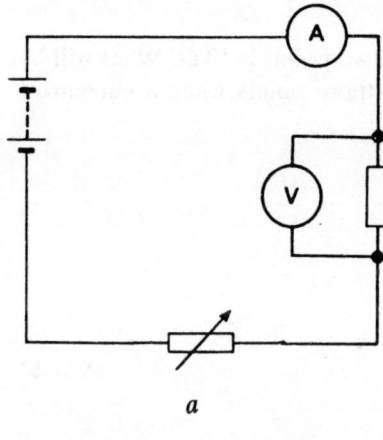

a

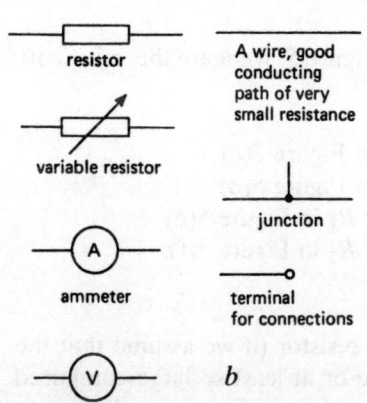

resistor

A wire, good
conducting
path of very
small resistance

variable resistor

junction

ammeter

terminal
for connections

voltmeter

b

Figure 3 (a) *A circuit diagram*
(b) *Symbols used in the diagram*

Figure 4 *A circuit with no branches,*
the current is the same at all points

This relationship is known as *Ohm's law*:

At a constant temperature, the potential difference is proportional to the current.

Electrical circuits are often represented by circuit diagrams. Figure 3 is an example of a simple circuit diagram. In such diagrams symbols are used to represent the various circuit components. The diagram also does not represent the layout of the circuit, the way it is assembled in a box or even on the bench, but the way by which the various components are connected together. In the example given, a wire connects one terminal of the d.c. source to one terminal of the ammeter. From the other terminal of the ammeter a wire connects it to a junction where wires from the voltmeter and the resistor meet. The other side of the resistor is again a junction, with a wire running from this junction to a variable resistor. The other side of the variable resistor is connected to the d.c. supply.

The direction indicated on a circuit diagram for the current is from the positive terminal of the d.c. supply towards the negative terminal.

Current, as we have earlier considered, can be thought of as being like water flowing along a trough. Suppose we have a water pump and pump water along a pipe that has no branches. At any position along the pipe the rate of water flow will be the same. This must be true regardless of whether the pipe changes its cross-section or has bends in it. The rate at which water enters the pipe equals the rate at which it leaves. The same is true with electrical current.

In an electrical circuit with no branches the current is the same at all points in the circuit.

Figure 4 illustrates this.

Example 1
In an electrical circuit the potential difference between two points was 2 V when the current between those points was 0.5 A. What is the electrical resistance between those points?

$V = 2$ V, $I = 0.5$ A

$$R = \frac{V}{I}$$

$$= \frac{2}{0.5} \qquad \frac{V}{A}$$

$$= 4 \ \Omega$$

Example 2

The electrical resistance between two points is 10 Ω. What will be the potential difference between those points when a current of 0.4 A passes between them?

$R = 10 \ \Omega, I = 0.4 \ \text{A}$

$$R = \frac{V}{I}$$

Therefore $V = IR$

$$= 0.4 \times 10 \qquad\qquad\qquad \text{A} \times \Omega$$

$$= 4 \ \text{V}$$

Example 3

For the circuit diagrams shown in Figure 5, what are the values of:
(a) The current I_1 in Figure 5(a)
(b) The current I_2 in Figure 5(b)
(c) The potential difference V_1 in Figure 5(c)
(d) The potential difference V_2 in Figure 5(d)
(e) The resistance of the resistor R_1 in Figure 5(e)
(f) The resistance of the resistor R_2 in Figure 5(f)

(a) $V = 2 \ \text{V}, R = 10 \ \Omega$

I_1 is the current through the 10 Ω resistor (if we assume that the voltmeter has an infinite resistance or at least so large compared with the resistor that all the current can be assumed to pass through the resistor).

$$R = \frac{V}{I}$$

Therefore $I_1 = \dfrac{V}{R}$

$$= \frac{2}{10} \qquad\qquad\qquad \frac{\text{V}}{\Omega}$$

$$= 0.2 \ \text{A}$$

(b) $V = 1 \ \text{V}, R = 5 \ \Omega$

I_2 is the current through the 5 Ω resistor (with the same assumption as in (a) above).

$$R = \frac{V}{I}$$

Figure 5

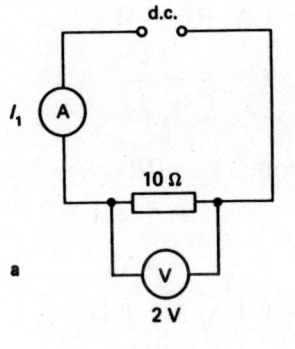

a

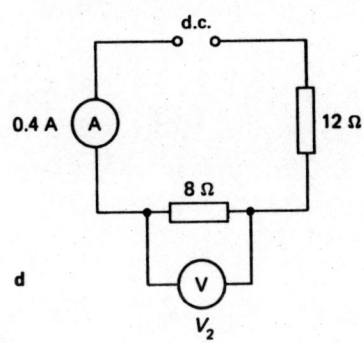

d

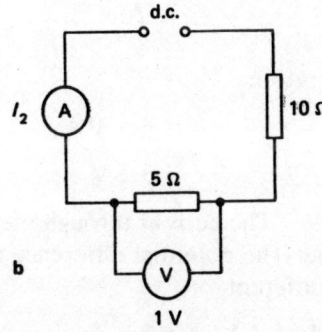

b

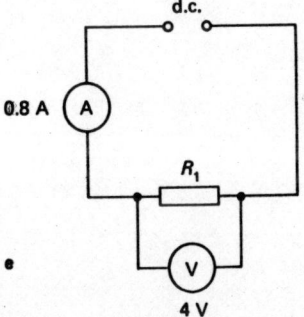

e

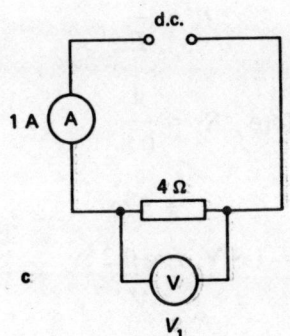

c

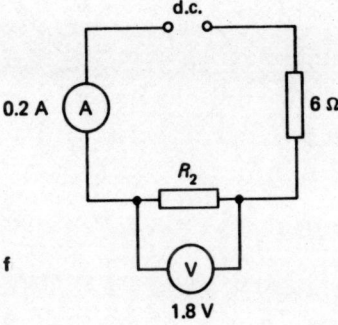

f

Therefore $\quad I_2 = \dfrac{V}{R}$

$\qquad\qquad = \dfrac{1}{5} \qquad\qquad\qquad\qquad \dfrac{V}{\Omega}$

$\qquad\qquad = 0.2\ A$

Note: The current through the 5 Ω and 10 Ω resistors will be the same. The potential differences across each resistor will, however, be different.

(c) $I = 1$ A, $R = 4\ \Omega$

$$R = \frac{V}{I}$$

Therefore $V_1 = IR$

$\qquad\qquad\quad = 1 \times 4$ A $\times$ Ω

$\qquad\qquad\quad = 4$ V

(d) $I = 0.4$ A, $R = 8\ \Omega$

$$R = \frac{V}{I}$$

Therefore $V_2 = RI$

$\qquad\qquad\quad = 8 \times 0.4$ A $\times$ Ω

$\qquad\qquad\quad = 3.2$ V

Note: The current through the 8 Ω and 12 Ω resistors will be the same. The potential difference across each resistor will, however, be different.

(e) $V = 4$ V, $I = 0.8$ A

$$R = \frac{V}{I}$$

Therefore $R_1 = \dfrac{4}{0.8}$ $\dfrac{V}{A}$

$\qquad\qquad\quad = 5\ \Omega$

(f) $V = 1.8$ V, $I = 0.2$ A

$$R = \frac{V}{I}$$

Therefore $R_2 = \dfrac{1.8}{0.2}$ $\dfrac{V}{A}$

$\qquad\qquad\quad = 9\ \Omega$

Example 4

Using the circuit shown in Figure 3, the current was varied by varying the resistance of the resistor and the following series of readings of current and potential difference obtained.

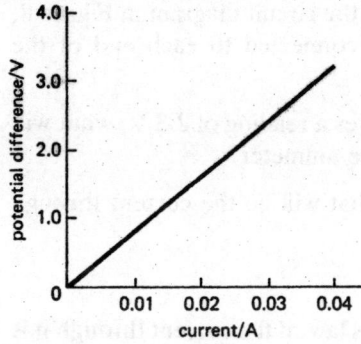

Figure 6

Potential difference (V)	0	0.8	1.6	2.4	3.2
Current (A)	0	0.01	0.02	0.03	0.04

(a) Plot a graph of potential difference against current.
(b) Does the resistor obey Ohm's law?
(c) What is the resistance of the resistor?

(a) Figure 6 shows the graph.
(b) The graph is a straight line passing through the origin, so Ohm's law is obeyed.
(c) The resistance is 3.2/0.04 = 80 Ω.

Example 5

The following are the potential difference/current data for a resistor.
(a) Plot the potential difference/current graph
(b) Does the resistor obey Ohm's law?
(c) What is its resistance at 0.2 A?

Potential difference (V)	0	0.5	1	1.5	2	2.5
Current (A)	0	0.1	0.2	0.3	0.42	0.57

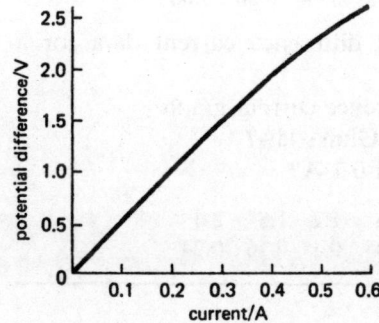

Figure 7

(a) Figure 7 shows the graph.
(b) Ohm's law is not obeyed for the entire range of current. The graph is only a straight line through the origin up to a current of about 0.3 A. Ohm's law is thus only obeyed up to about 0.3 A.
(c) For a current of 0.2 A, the potential difference is 1 V and the current 0.2 A, therefore $R = V/I = 1/0.2 = 5\ \Omega$.

Note: Because the graph is not a straight line over the entire current range, the resistance is not constant over the entire current range. The resistance decreases when the current is greater than 0.3 A.

Self-assessment questions

1   In an electrical circuit the potential difference between two points was 1.2 V when the current between those points was 1 A. What is the electrical resistance between those points?

2   If the electrical resistance between two points is 20 Ω, what will be the potential difference between those points when a current of 0.5 A passes between them?

3   When the potential difference across a resistor of resistance 40 Ω is 0.2 V, what is the current passing through the resistor?

4   What is the resistance of a resistor if a current of 0.12 A through that resistor produces a potential difference of 3.6 V across it?

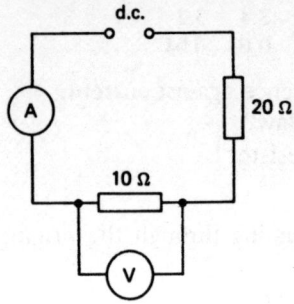

Figure 8

5   For the circuit represented by the circuit diagram in Figure 8, what circuit components are connected to each end of the 10 Ω resistor?

6   If the voltmeter in Figure 8 gives a reading of 2.3 V, what will be the current indicated by the ammeter?

7   For the circuit in Figure 8, what will be the current through the 20 Ω resistor?

8   State Ohm's law.

9   For a resistor that obeys Ohm's law, if the current through it is trebled, how does the potential difference change?

10   The following are potential difference and current data for a resistor.
   (a)   Plot the potential difference current graph
   (b)   Does the resistor obey Ohm's law?
   (c)   What is its resistance?

Potential difference (V)	0	0.5	1	1.5	2
Current (A)		0	0.02	0.04	0.06	0.08

11   The following are potential difference current data for a resistor.
   (a)   Plot the potential difference current graph
   (b)   Does the resistor obey Ohm's law?
   (c)   What is its resistance at 0.1 A?

Potential difference (V)	0	0.6	1.2	1.8	2.4
Current (A)		0	0.05	0.1	0.16	0.24

Multiples of units

While it is always possible to quote currents in amps, sometimes the current may be so large or so small that quoting it that way is awkward. Thus, to take an example, suppose we have a current of 0.001 A. We could write this as 1×10^{-3} A or, alternatively, we could prefix the unit by a term to indicate this multiplying factor of 10^{-3} (*Note*: 10^{-3} is 1/1000). The term we use in this case is milli (symbol m). Thus, the current can be written as 1 mA (milliamp).

To take another example, we could have a current of 3000 A. This could be written as 3×10^3 A (*Note*: 10^3 is 1000) or prefix the unit by a term to indicate this multiplying factor of 10^3. The term used is kilo (symbol k). Thus, the current can be written as 1 kA (kiloamp).

Table 1 shows the multiples that are used.

Table 1

Multiplication factor		Prefix	Prefix symbol
10^{-12} or $\dfrac{1}{1\,000\,000\,000\,000}$		pico	p
10^{-9} or $\dfrac{1}{1\,000\,000\,000}$		nano	n
10^{-6} or $\dfrac{1}{1\,000\,000}$		micro	μ
10^{-3} or $\dfrac{1}{1000}$		milli	m
10^{3}	or 1000	kilo	k
10^{6}	or 1 000 000	mega	M
10^{9}	or 1 000 000 000	giga	G
10^{12}	or 1 000 000 000 000	tera	T

These prefixes to the units are not restricted to current or even just electrical quantities but are used in all scientific quantities.

Example 6
Write 2500 A in kA.
$$2500 \quad = 2.5 \times 1000 = 2.5 \times 10^3$$
Therefore   2500 A = 2.5 kA

Example 7
Write 0.035 A in mA.
$$0.035 \quad = 35 \times 10^{-3} = \frac{35}{1000}$$

Therefore   0.035 A = 35 mA

Example 8
Write 120 mA in A.

milli is a multiplying factor of $10^{-3} = \dfrac{1}{1000}$

Therefore $120\ \text{mA} = 120 \times 10^{-3} = \dfrac{120}{1000} = 0.120\ \text{A}$

Example 9
Write 0.000 120 V in μV.
$$0.000\ 120 \quad = 120 \times 10^{-6} = \frac{120}{1\,000\,000}$$

Therefore   0.000 120 V = 120 μV

Example 10

Write 30 MΩ in Ω.

mega is a multiplying factor of $10^6 = 1\ 000\ 000$.

Therefore $30\ \text{M}\Omega = 30 \times 10^6 = 30 \times 1\ 000\ 000 = 30\ 000\ 000\ \Omega$

Example 11

The potential difference across a resistor was 1.2 V when the current through it was 60 mA. What is the resistance of the resistor?

$V = 1.2\ \text{V}, I = 60\ \text{mA} = 60 \times 10^{-3}\ \text{A}$

$$R = \frac{V}{I}$$

$$= \frac{1.2}{60 \times 10^{-3}} \qquad\qquad \frac{\text{V}}{\text{A}}$$

$$= 20\ \Omega$$

Example 12

What is the potential difference across a 2 MΩ resistor when a current of 4 mA passes through it?

$R = 2\ \text{M}\Omega = 2 \times 10^6 \Omega, I = 4\ \text{mA} = 4 \times 10^{-3}\text{A}$

$$R = \frac{V}{I}$$

Therefore $V = IR$

$\qquad\qquad = 4 \times 10^{-3} \times 2 \times 10^6 \qquad\qquad \text{A} \times \Omega$

$\qquad\qquad = 8 \times 10^3\ \text{V}$

$\qquad\qquad = 8\ \text{kV}$

Self-assessment questions

12 Write:
  (a) 0.050 A in mA
  (b) 200 A in kA
  (c) 0.000 400 A in μA
  (d) 20 mA in A
  (e) 3 kA in A
  (f) 0.3 mA in μA
  (g) 2000 V in kV
  (h) 0.000 100 V in μV
  (i) 15 mV in V
  (j) 3000 μV in V

(k)   2500 Ω in kΩ
(l)   140 000 Ω in MΩ
(m)   4 MΩ in Ω
(n)   5 kΩ in Ω

13   The potential difference across a resistor was 0.6 V when the current through it was 20 mV. What is the resistance of the resistor?

14   What is the potential difference across a 0.5 MΩ resistor when the current through it is 20 μA?

15   The potential difference across a resistor of resistance 2.0 kΩ is 4.0 V. What is the current through the resistor?

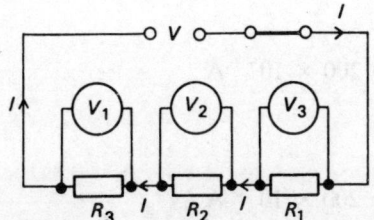

Figure 9   *A series circuit*

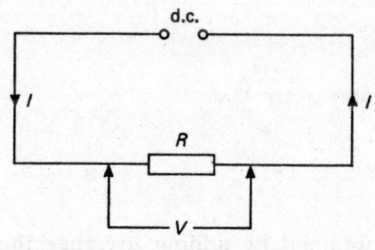

Figure 10   *The equivalent circuit*

Series circuits

A series circuit is where all the components are connected in one long line. Figure 9 shows such a circuit. With a series circuit the current through each component is the same. The potential differences across each component may not however be the same, i.e. V_1, V_2 and V_3 may all be different:

For resistor R_1 the potential difference $V_1 = IR_1$
For resistor R_2 the potential difference $V_2 = IR_2$
For resistor R_3 the potential difference $V_3 = IR_3$

Suppose we want to find the equivalent circuit (see Figure 10) which takes the same current I and has the same overall potential difference, i.e. $V = V_1 + V_2 + V_3$.

For resistor R the potential difference $V = IR$.

And $V = V_1 + V_2 + V_3$

So $IR = IR_1 + IR_2 + IR_3$

$$\boxed{R = R_1 + R_2 + R_3}$$

Thus we can replace the three resistors by a single resistor of resistance R, provided it equals the sum of the three resistances, and have exactly the same effect on the circuit current and potential difference.

An important point to note is that *in a series circuit all components carry the same current and their individual potential differences are added together to give the total potential difference.*

Example 13

Three resistors are connected in series. If they have resistances of
5 Ω, 10 Ω and 20 Ω what is the value of the total resistance?
$R_1 = 5\ \Omega, \quad R_2 = 10\ \Omega, \quad R_3 = 20\ \Omega$

$R = R_1 + R_2 + R_3$

$\quad = 5 + 10 + 20$

$\quad = 35\ \Omega$

Example 14

Three resistors are connected in series. If they have resistances of
10 Ω, 15 Ω and 20 Ω and a current of 200 mA passes through
them, what is:

(a)   The potential difference across each resistor
(b)   The value of the total resistance
(c)   The potential difference across the total resistance?

(a) $R_1 = 10\ \Omega, I = 200\ \text{mA} = 200 \times 10^{-3}\ \text{A}$
       $V = IR$
       $\quad = 200 \times 10^{-3} \times 10$
       $\quad = 2\ \text{V}$
       $R_2 = 15\ \Omega, I = 200\ \text{mA} = 200 \times 10^{-3}\ \text{A}$
       $V = IR$
       $\quad = 200 \times 10^{-3} \times 15$
       $\quad = 3\ \text{V}$
       $R_3 = 20\ \Omega, I = 200\ \text{mA} = 200 \times 10^{-3}\ \text{A}$
       $V = IR$
       $\quad = 200 \times 10^{-3} \times 20$
       $\quad = 4\ \text{V}$

(b) $R_1 = 10\ \Omega, R_2 = 15\ \Omega, R_3 = 20\ \Omega$
       $R = R_1 + R_2 + R_3$
       $\quad = 10 + 15 + 20$
       $\quad = 45\ \Omega$

(c) $R = 45\ \Omega, I = 200\ \text{mA} = 200 \times 10^{-3}\ \text{A}$
       $V = IR$
       $\quad = 200 \times 10^{-3} \times 45$
       $\quad = 9\ \text{V}$

This answer could have been obtained by adding together the
potential differences across each of the three resistors, i.e. 2 + 3 +
4 = 9 V.

Self-assessment questions

16   Two resistors are connected in series. If they have resistances
of 20 Ω and 40 Ω what is the value of the total resistance?

17 Three resistors are connected in series. If they have resistances of 1 kΩ, 5 kΩ and 10 kΩ what is the value of the total resistance?

18 Three resistors are connected in series. If they have resistances of 10 Ω, 20 Ω and 100 Ω and a current of 20 mA passes through them, what is:
(a) The potential difference across each resistor
(b) The value of the total resistance
(c) The potential difference across the total resistance?

19 A resistance of 80 Ω is required in a circuit, but only resistors of value 10 Ω and 50 Ω are available. How could you obtain the required resistance by a series connection of resistors from the values given? You may assume that more than one resistor is available at each value.

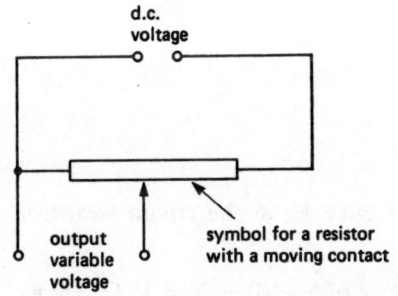

Figure 11 *A potential divider circuit*

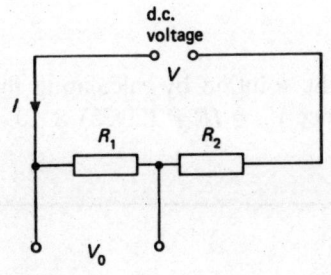

Figure 12

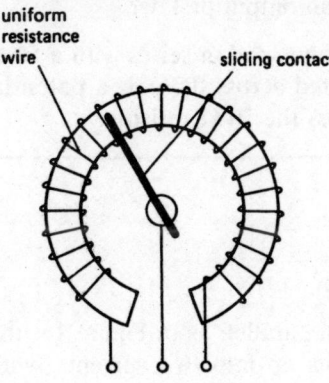

Figure 13 *A potentiometer*

A potential divider

A series circuit can be used to subdivide a potential difference. An example of this is with a power pack which supplies a variable voltage. Within the power pack a series circuit is used to subdivide the available voltage. Figure 11 shows the type of circuit used. A variable resistor is used with the full d.c. voltage being connected across it. The variable voltage is tapped off by connections to one end of the variable resistor and to a moving contact. By moving this contact the resistance between the output terminals is varied. This has the effect of varying the potential difference between those terminals and hence the output voltage.

Figure 12 shows the effective arrangement of Figure 11 when the slider is at some particular position. The total resistance of the variable resistor is subdivided so that R_1 is between the output terminals and R_2 outside those terminals. The output potential difference $V_o = IR_1$. But the total voltage $V = I(R_1 + R_2)$. Hence:

$$I = \frac{V}{R_1 + R_2}$$

and so $V_o = IR_1 = V\left(\frac{R_1}{R_1 + R_2}\right)$

As $R_1 + R_2$ is the total resistance, then $R_1/(R_1 + R_2)$ is the fraction of the total resistance that is between the output terminals.

A variable resistor with a sliding contact, i.e. a three terminal device, is generally known as a potentiometer. Figure 13 shows the form such a potentiometer can take.

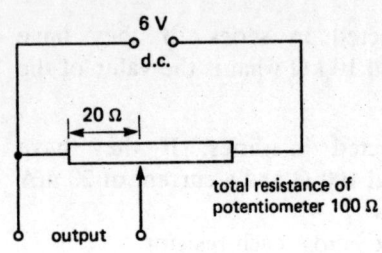

Figure 14

Example 15

What will be the output voltage for the arrangement shown in Figure 14?

Using the symbols used in Figure 12, $R_1 = 20\ \Omega$, $R_1 + R_2 = 100\ \Omega$, $V = 6\ V$.

$$V_o = V\left(\frac{R_1}{R_1 + R_2}\right)$$

$$= 6 \times \frac{20}{100}$$

$$= 1.2\ V$$

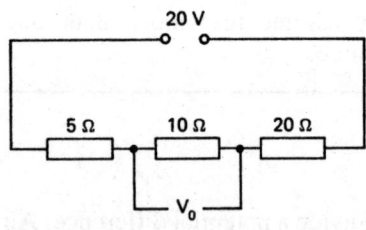

Figure 15

Example 16

What will be the potential difference V_o in the circuit shown in Figure 15?

The 20 V is across a total resistance of $5 + 10 + 20 = 35\ \Omega$. Hence the fraction of this which is across the 10 Ω resistor is:

$$V_o = 20 \times \frac{10}{35}$$

$$= 5.7\ V$$

Note: You could have obtained the solution by calculating the circuit current, $I = 20/35$ A and hence $V_o = IR = (20/35) \times 10 = 5.7$ V.

Self-assessment questions

20   A 12 V d.c. power supply is to be converted into a 0–12 V variable power supply by using a potential divider circuit. If the potentiometer has a resistance of 100 Ω, what will be the resistance of that part of the potentiometer across which the output is taken when there is an output of 1 V?

21   What size resistor should be connected in series with a 50 Ω resistor if when 6 V is connected across the pair a potential difference of 2 V appears across the 50 Ω resistor?

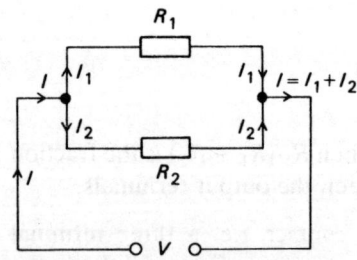

Figure 16   *A parallel circuit*

Parallel circuits

When components are connected in parallel, as in Figure 16, the current I from the d.c. source splits up into two currents, with current I_1 passing through resistor R_1 and current I_2 through resistor R_2. Later these two currents recombine to give the original current I.

$$I = I_1 + I_2$$

Since the ends of both the resistors are connected together the potential difference across each resistor must be the same.

For resistor R_1, the potential difference $V = I_1R_1$.
For resistor R_2, the potential difference $V = I_2R_2$.

Hence $I_1 = V/R_1$ and $I_2 = V/R_2$, and so:

$$I = I_1 + I_2$$

$$I = \frac{V}{R_1} + \frac{V}{R_2}$$

$$I = V\left(\frac{1}{R_1} + \frac{1}{R_2}\right)$$

If we replace the parallel arrangement of resistors by a single equivalent resistor then it must have a potential difference of V across it and a current I through it. It will therefore have a resistance R, where:

$$R = \frac{V}{I}$$

Rearranging the equation for the parallel resistors gives:

$$\frac{I}{V} = \frac{1}{R_1} + \frac{1}{R_2}$$

Hence:
$$\boxed{\frac{1}{R} = \frac{1}{R_1} + \frac{1}{R_2}}$$

An important point to note is that *in a parallel circuit all the components have the same potential difference and their currents are added together to give the total current.*

Example 17

A 10 Ω resistor and a 5 Ω resistor are connected in parallel. What is the total resistance of the combination?

$$R_1 = 10 \ \Omega, \ R_2 = 5 \ \Omega$$

$$\frac{1}{R} = \frac{1}{R_1} + \frac{1}{R_2}$$

$$\frac{1}{R} = \frac{1}{10} + \frac{1}{5}$$

$$\frac{1}{R} = \frac{1+2}{10}$$

$$R = \frac{10}{3}$$

$$R = 3\tfrac{1}{3}\ \Omega$$

Example 18
Two 100 Ω resistors are connected in parallel. What is the total resistance of the combination?

$$R_1 = 100\ \Omega,\ R_2 = 100\ \Omega$$

$$\frac{1}{R} = \frac{1}{R_1} + \frac{1}{R_2}$$

$$\frac{1}{R} = \frac{1}{100} + \frac{1}{100}$$

$$\frac{1}{R} = \frac{2}{100}$$

$$R = 50\ \Omega$$

Example 19
A 40 Ω resistor and a 20 Ω resistor are connected in parallel. If a d.c. source supplies a current of 0.2 A to the combination, what is:

(a)   The total resistance of the combination
(b)   The potential difference across the resistors
(c)   The current through each resistor?

The circuit can be considered to be like that in Figure 16 with $I = 0.2$ A, $R_1 = 40\ \Omega$ and $I_2 = 20\ \Omega$.

(a)
$$\frac{1}{R} = \frac{1}{R_1} + \frac{1}{R_2}$$

$$\frac{1}{R} = \frac{1}{40} + \frac{1}{20}$$

$$\frac{1}{R} = \frac{1+2}{40}$$

$$R = \frac{40}{3}$$

$$R = 13\tfrac{1}{3}\ \Omega$$

(b)   As $V = IR$
$$V = 0.2 \times 13\tfrac{1}{3}$$
$$= 2.7 \text{ V}$$

(c)   For resistor R_1:
$$V = I_1 R_1$$
$$I_1 = \frac{V}{R_1} = \frac{2.7}{40}$$

$$= 0.067 \text{ A}$$

For resistor R_2:
$$V = I_2 R_2$$
$$I_2 = \frac{V}{R_2} = \frac{2.7}{20}$$

$$= 0.135 \text{ A}$$

We can check these answers because we know that the combined current must equal 0.2 A. Because of the rounding of the data during the calculation this is only approximately true.

Self-assessment questions

22   A 20 Ω and a 10 Ω resistor are connected in parallel. What is the total resistance of the combination?

23   Three 12 Ω resistors are connected in parallel. What is the total resistance of the combination?

24   A 100 Ω resistor and a 40 Ω resistor are connected in parallel. If a d.c. source supplies a current of 10 mA to the combination, what is:
(a)   The total resistance of the combination
(b)   The potential difference across the resistors
(c)   The current through each resistor?

25   A 2 kΩ and a 1 kΩ resistor are connected in parallel and the combination draws a current of 20 mA from the d.c. supply. What is:
(a)   The potential difference across the parallel circuit
(b)   The current taken by each resistor?

26   A resistor R is connected in parallel with a 50 Ω resistor in order to reduce the resistance of the combination to 5 Ω. What must be the resistance of R?

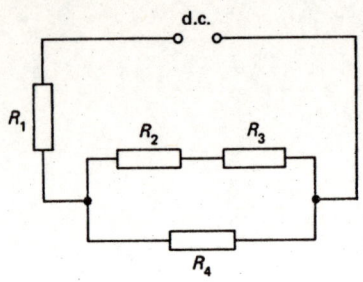

Figure 17   *A circuit with series and parallel resistors*

Series-parallel circuits

Many circuits cannot be described as just series or just parallel circuits but can be described as combinations of series and parallel arrangements of circuit components. Figure 17 shows such an arrangement. The way to tackle such a situation is to reduce the series-parallel circuit to a number of just series and just parallel circuits and solve each one in turn.

Figure 18 illustrates how the circuit in Figure 17 can be reduced to just series and just parallel circuits. Thus resistors R_2 and R_3 are a series arrangement which can be replaced by a single equivalent resistor R_5, where:

$$R_5 = R_2 + R_3$$

This R_5 resistor can then be considered in parallel with resistor R_4 and replaced with a single resistor R_6, where:

$$\frac{1}{R_6} = \frac{1}{R_5} + \frac{1}{R_4}$$

This R_6 resistor can then be considered in series with the R_1 resistor and replaced with a single resistor R_7, where:

$$R_7 = R_6 + R_1$$

Example 20
Calculate the total resistance of the circuit shown in Figure 19.

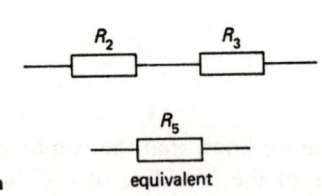

a   equivalent

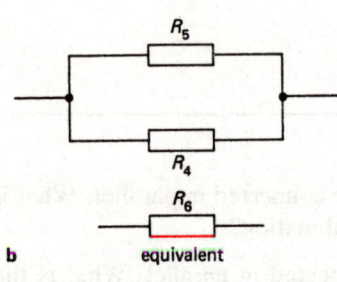

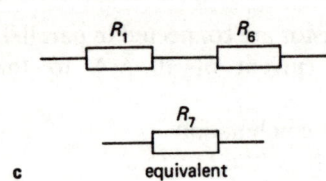

b   equivalent

c   equivalent

Figure 18

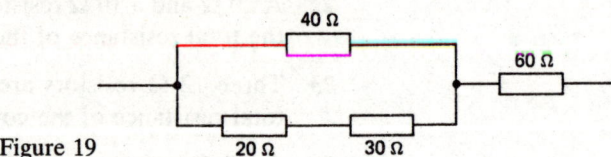

Figure 19

Consider the series combination of the 20 Ω and the 30 Ω resistors. They can be replaced by a single resistor having the value:

$$R = R_1 + R_2$$
$$= 20 + 30$$
$$= 50 \ \Omega$$

The resulting circuit is now as in Figure 20.

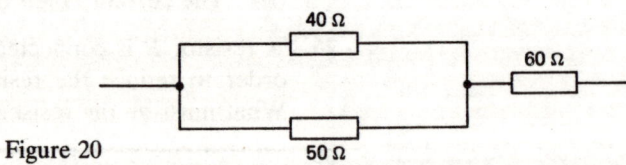

Figure 20

Now consider the parallel arrangement of the 40 Ω and the 50 Ω resistors. These can be replaced by a single resistor having the value R, where:

$$\frac{1}{R} = \frac{1}{R_3} + \frac{1}{R_4}$$

$$\frac{1}{R} = \frac{1}{40} + \frac{1}{50}$$

$$\frac{1}{R} = \frac{5 + 4}{200}$$

$$R = \frac{200}{9}$$

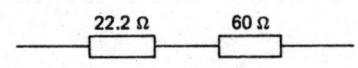

$$R = 22.2 \ \Omega$$

Figure 21

The resulting circuit is now as in Figure 21.

Now consider the series arrangement of the 22.2 Ω and the 60 Ω resistors. These can be replaced by a single resistor having the value:

$$R = R_5 + R_6$$
$$= 22.2 + 60$$
$$= 82.2 \ \Omega$$

This is the total resistance of the circuit.

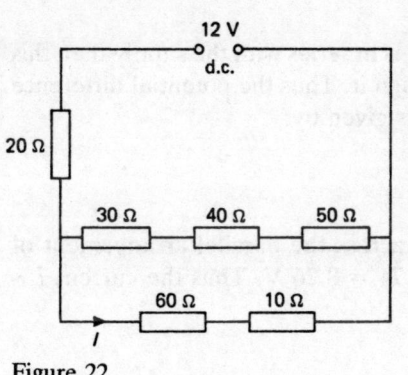

Figure 22

Example 21
Calculate:
(a) The total resistance of the circuit shown in Figure 22
(b) The potential difference across the 20 Ω resistor
(c) The current I

(a) For the series combination of 60 Ω and 10 Ω resistors, the combined resistance is:

$$R = R_1 + R_2$$
$$= 60 + 10$$
$$= 70 \ \Omega$$

For the series combination of 30 Ω, 40 Ω and 50 Ω resistors, the combined resistance is:

$$R = R_3 + R_4 + R_5$$
$$= 30 + 40 + 50$$
$$= 120 \ \Omega$$

Now considering the parallel arrangement of 70 Ω and 120 Ω, the combined resistance is:

$$\frac{1}{R} = \frac{1}{R_6} + \frac{1}{R_7}$$

$$= \frac{1}{70} + \frac{1}{120}$$

$$= \frac{120 + 70}{70 \times 120}$$

$$R = \frac{70 \times 120}{190}$$

$$= 44.2 \ \Omega$$

Now combining this resistance with that of the 20 Ω resistor gives

$$R = R_8 + R_9$$
$$= 44.2 + 20$$
$$= 64.2 \ \Omega$$

(b) The current taken from the source is given by $V = IR$, where V is the source potential difference and R the total circuit resistance. Thus:

$$I = \frac{V}{R} = \frac{12}{64.2}$$

$$= 0.187 \ \text{A} \ (187 \ \text{mA})$$

Because the 20 Ω resistor is in series with the supply then this must be the current through it. Thus the potential difference across the 20 Ω resistor is given by:

$$V = IR = 0.187 \times 20$$
$$= 3.74 \ \text{V}$$

(c) The potential difference across the parallel arrangement of resistors must be $12 - 3.74 = 8.26$ V. Thus the current I is given by:

$$I = \frac{V}{R} = \frac{8.26}{70}$$

Note that the resistance value used is that of the total resistance through which the current I is flowing in that arm of the circuit. Hence:

$$I = 0.118 \ \text{A} \ (118 \ \text{mA})$$

Self-assessment questions

27 Calculate the total resistances of the circuits shown in Figure 23.

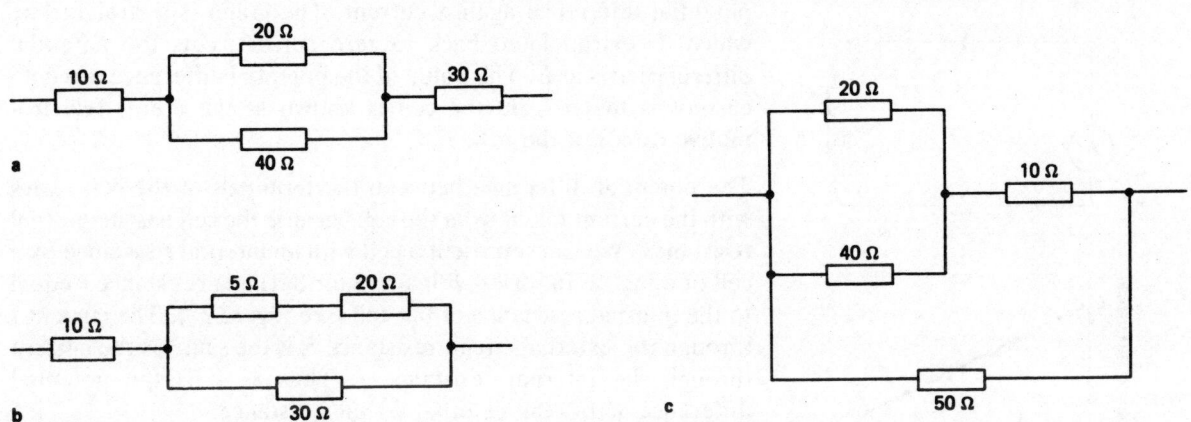

Figure 23

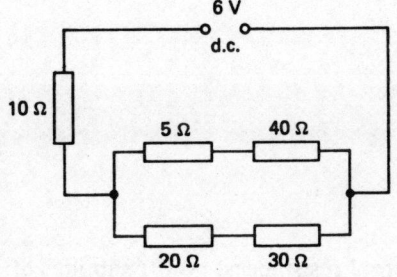

Figure 24

28 For the circuit shown in Figure 24 calculate:
(a) The total resistance
(b) The current through the 10 Ω resistor
(c) The current through the 20 Ω resistor.

29 For the circuit shown in Figure 25 calculate:
(a) The current *I*
(b) The potential difference between points *A* and *B*.

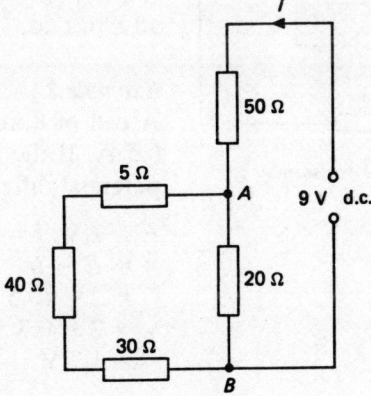

Figure 25

symbol for cell

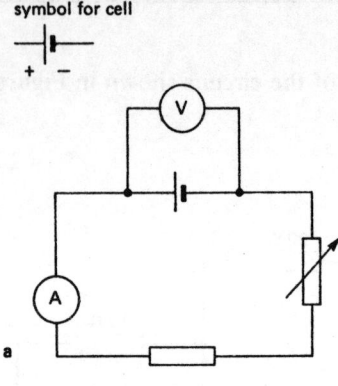

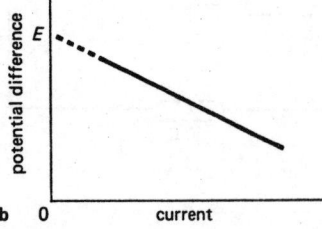

b   0   current

Figure 26

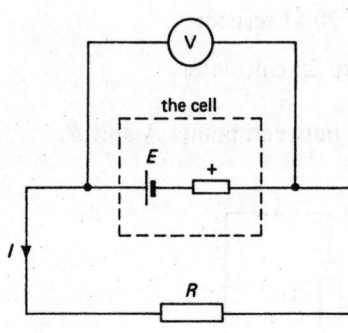

Figure 27   *Internal resistance*

Electromotive-force (e.m.f.)

Figure 26 shows a simple circuit involving a d.c. source, in this case a cell. The potential difference across the cell is measured for different currents taken from the cell and a graph plotted of potential difference against current. The graph is a straight line which, if extrapolated back to zero current, cuts the potential difference axis at E. This value of the potential difference when no current is taken from the cell is known as the e.m.f. (electro-motive-force) of the cell.

The potential difference between the terminals of the cell varies with the current taken from the cell because the cell has an internal resistance. We can represent a cell with an internal resistance by a cell of e.m.f. E in series with a resistor having a resistance r equal to the internal resistance of the cell (see Figure 27). The current I through the external circuit resistance R is the same as the current through the internal resistance r. Thus as V is the potential difference across the external circuit resistance:

$$V = IR$$

But the circuit e.m.f. is applied to a series combination of r and R. Hence:

$$E = I(R + r)$$
$$E = IR + Ir$$
$$E = V + Ir$$

$$\boxed{V = E - Ir}$$

When I is zero then $V = E$.

It is not only cells that have internal resistances, power supplies of all types do.

Example 22
A cell of e.m.f. 2 V delivers a current to an external circuit of 0.5 A. If the cell has an internal resistance of 0.6 Ω what is the potential difference across the cell terminals?

$$E = 2 \text{ V}, I = 0.5 \text{ A}, r = 0.6 \ \Omega$$
$$\begin{aligned} V &= E - Ir \\ &= 2 - (0.5 \times 0.6) \\ &= 2 - 0.3 \\ &= 1.7 \text{ V} \end{aligned}$$

Self-assessment questions

30   A cell of e.m.f. 2.2 V delivers a current to an external circuit of 0.8 A. If the potential difference between the terminals of the cell is found to be 1.5 V, what is the internal resistance of the cell?

31 A cell of e.m.f. 2.1 V and internal resistance 0.5 Ω delivers a current of 0.2 A to an external circuit. What is the potential difference between the terminals of the cell?

Kirchhoff's laws

These laws can be expressed as:

First law: *The sum of the current flowing towards any point in a circuit is equal to the sum of the current flowing away from it.*

Second law: *Around any closed circuit loop, the algebraic sum of the e.m.f.s and the potential differences is zero.*

To illustrate the first law, consider the circuit junction shown in Figure 28. The current entering the junction is I_1 with the current I_2 and current I_3 leaving the junction. Hence: $I_1 = I_2 + I_3$

If current is thought of as rather like a flow of water, this first law can be interpreted as saying that the rate of flow of water to a junction must equal the rate at which it flows away from the junction. Water cannot accumulate at the junction.

To illustrate the second law, consider the closed circuit loop shown in Figure 29. In applying this law the mathematical signs attached to the potential differences and e.m.f.s are very important. The term 'algebraic' in the law means that the sum is taken with due regard to the sign. The following is a general procedure to be adopted with any closed loop.

1 Decide on a direction for the current and indicate this, all the way round the loop, by arrows.
2 For each source of e.m.f. put an arrow pointing from the negative terminal to the positive terminal. This is taken as the direction of the e.m.f. of the source.
3 For each resistor put an arrow pointing in the opposite direction to the current through the resistor concerned. This is taken as the direction of the potential difference, *IR*, of the resistor.
4 Decide on one particular point on the circuit loop and proceed round the loop writing down the potential differences and e.m.f.s, with due regard to sign. Continue round the loop until you end up at the point you started. If you cannot end at the point you started from then it is not a closed loop and you cannot apply the law.

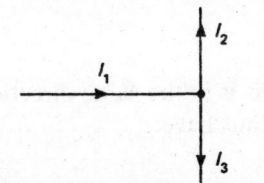

Figure 28

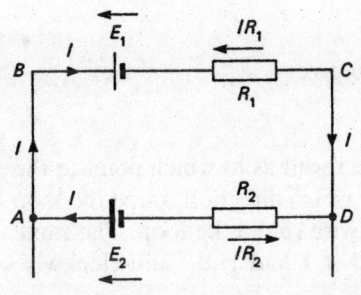

Figure 29

Starting, for the loop in Figure 29, at point A and proceeding to B, no potential difference or source of e.m.f. is encountered, so there is nothing to write down. Going from B to C we first encounter E_1. This has an arrow in the opposite direction to which we are moving round the circuit and so is negative. We thus have:

$$- E_1$$

The next term we meet in going from B to C is IR_1. This is also in the opposite direction to which we are going and so is negative. We now have:

$$- E_1 - IR_1$$

In going from C to D we encounter no source of e.m.f. or potential difference and so there are no terms to add for that part of the loop. In going from D to A we meet a potential difference IR_2 which is in the opposite direction to which we are going and so is negative. We now have:

$$- E_1 - IR_1 - IR_2$$

We next encounter the e.m.f. E_2. This is in the same direction as we are going and so is positive. We thus have:

$$- E_1 - IR_1 - IR_2 + E_2$$

We are now back at the point A we started from. Hence, applying the second law means that the above sum must equal zero.

$$- E_1 - IR_1 - IR_2 + E_2 = 0$$

This equation can be rearranged to give:

$$E_2 - E_1 = IR_1 + IR_2$$

It makes no difference to the above result as to which point in the loop we start to proceed round it or which direction round the loop we go. In the example I went clockwise round the loop. The same answer would have been obtained if I had gone anti-clockwise round the loop.

Figure 30 shows a circuit with two closed circuit loops. For each of these loops we can arrive at an equation in the same way as above.

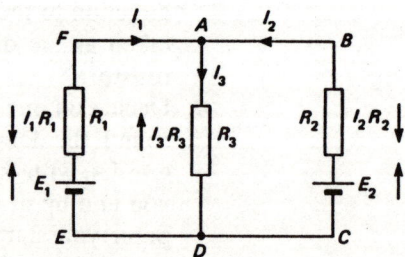

Figure 30

The current direction for each loop is determined and indicated. For that part of the circuit which is common to both loops (part AD) we can consider the current to be I_3. But as, according to the first law, the current entering a junction must equal the current leaving it, we have:

$$I_3 = I_1 + I_2$$

In order to keep the number of unknown currents to a minimum we use $I_1 + I_2$ in place of I_3 so that there are only two unknown currents appearing in the equations.

Taking loop $ABCD$, we can write:

$$I_2R_2 - E_2 + (I_1 + I_2)R_3 = 0$$

For the loop $ADEF$, we can write:

$$- (I_1 + I_2)R_3 + E_1 - I_1R_1 = 0$$

Generally in problems we know the resistances and the e.m.f.s, but not the currents. We therefore have two equations and two unknowns, i.e. I_1 and I_2. The equations can then be solved to give us the values of I_1 and I_2.

Example 23

If for the circuit shown in Figure 30 we have $E_1 = 2$ V, $R_1 = 50\ \Omega$, $R_2 = 100\ \Omega$, $E_2 = 4$ V and $R_3 = 120\ \Omega$, what will be the currents I_1 and I_2?

Using the equation for the loop $ABCD$:

$$I_2R_2 - E_2 + (I_1 + I_2)R_3 = 0$$
$$50I_2 - 4 + (I_1 + I_2) \times 120 = 0$$

Rearranging this equation gives:

$$120I_1 + 170I_2 - 4 = 0 \qquad\qquad \mathbf{1}$$

Using the equation for the loop $ADEF$:

$$- (I_1 + I_2)R_3 + E_1 - I_1R_1 = 0$$
$$- (I_1 + I_2) \times 120 + 2 - 50I_1 = 0$$

Rearranging this equation gives:

$$- 170I_1 - 120I_2 + 2 = 0 \qquad\qquad \mathbf{2}$$

We need to eliminate, by the use of equations **1** and **2**, one of the currents so that we can solve the other one. Rearranging equation **1**:

$$I_1 = \frac{4 - 170I_2}{120}$$

Rearranging equation **2**:

$$I_1 = \frac{2 - 120I_2}{170}$$

Hence:

$$\frac{4 - 170I_2}{120} = \frac{2 - 120I_2}{170}$$

$(4 \times 170) - (170 \times 170I_2) = (2 \times 120) - (120 \times 120I_2)$
$(4 \times 170) - (2 \times 120) = (170 \times 170I_2) - (120 \times 120I_2)$

$$I_2 = \frac{4 \times 170 - 2 \times 120}{170 \times 170 - 120 \times 120}$$

$$= 0.03 \text{ A } (30 \text{ mA})$$

Putting this value into equation **3** gives:

$$I_1 = \frac{4 - 170 \times 0.03}{120}$$

$$= -0.0092 \text{ A } (9.2 \text{ mA})$$

The minus sign means that the current is in the opposite direction to that adopted in Figure 30.

Note: If the direction of current I_1 is not what you would have expected from a consideration of the polarity of cell E_1, it is because the cell E_2 is driving a current not only through R_2 but also through R_1 and E_1.

Example 24

Calculate the current through the 10 Ω resistor in the circuit shown in Figure 31.

Using the first law:

$$I_3 = I_2 - I_1$$

Using the second law for the loop *ABEF*:

$$-3 + 5(I_2 - I_1) - 8I_1 + 2 = 0$$

This rearranges to give:

$$5I_2 - 13I_1 - 1 = 0 \qquad \qquad \textbf{1}$$

Using the second law for the loop *BCDE*:

$$-10I_2 - 5(I_2 - I_1) + 3 = 0$$

This rearranges to give:

$$-15I_2 + 5I_1 + 3 = 0 \qquad \qquad \textbf{2}$$

What is required is the current I_2, so we must eliminate I_1.

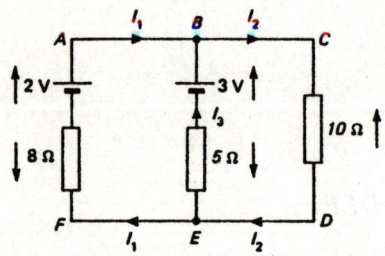

Figure 31

Rearranging **1** gives:

$$I_1 = \frac{5I_2 - 1}{13}$$

Rearranging **2** gives:

$$I_1 = \frac{15I_2 - 3}{5}$$

Hence $\dfrac{5I_2 - 1}{13} = \dfrac{15I_2 - 3}{5}$

$$25I_2 - 5 = 195I_2 - 39$$

$$I_2 = \frac{34}{170}$$

$$= 0.2 \text{ A (200 mA)}$$

Example 25

What is the potential difference across the 10 Ω resistor in Figure 32?

This question is a continuation of the previous example. In that example the current through the resistor was found to be 0.2 A. We thus have $I = 0.2$ A, $R = 10$ Ω, hence:

$$V = IR$$
$$= 0.2 \times 10$$
$$= 2 \text{ V}$$

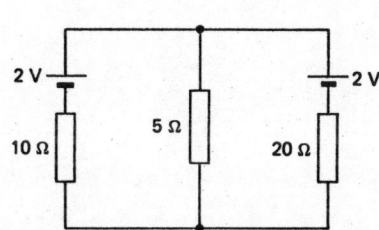

Figure 32

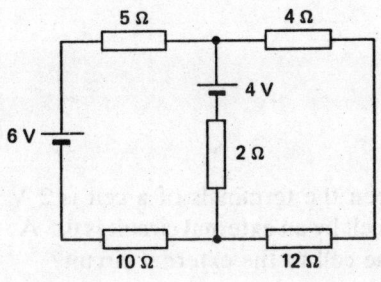

Figure 33

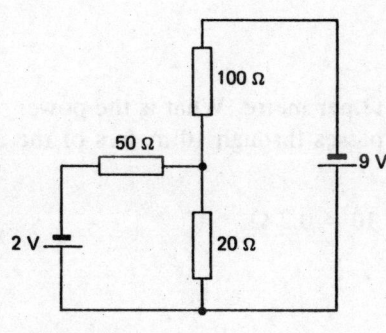

Figure 34

Self-assessment questions

32 Calculate the current through the 20 Ω resistor in the circuit in Figure 32.

33 Calculate the current through the 12 Ω resistor in the circuit in Figure 33.

34 Two cells are connected in parallel with each other and with a 5 Ω resistor. If one cell has an e.m.f. of 2 V and an internal resistance of 0.8 Ω and the other an e.m.f. of 4 V and an internal resistance of 1.8 Ω, what is the current through the 5 Ω resistor?

35 For the circuit shown in Figure 34, calculate the potential difference across the 20 Ω resistor.

Power in a d.c. circuit

Power is the rate at which energy is transferred or transformed from one form to another.

In the case of an electric current passing through a resistor the electrical energy is transformed to heat energy. For a current I moving through a potential difference V the electrical power P is IV. When I is in amps and V in volts then the power is in watts (W), where 1 watt is a transformation of 1 joule (J) of energy per second.

$$P = IV$$

For a resistor obeying Ohm's law we have $V = IR$, hence the power for such a resistor is:

$$P = IV = I(IR) = I^2R$$

or alternatively:

$$\boxed{P = IV = (V/R)V = V^2/R}$$

Example 26

What is the power taken from a d.c. source when it drives a current of 2 A through a resistance of 10 Ω?

$I = 2\ \text{A},\ R = 10\ \Omega$
$P = I^2R$
$\quad = 2^2 \times 10$
$\quad = 40\ \text{W}$

Example 27

If the potential difference between the terminals of a cell is 2 V when the current taken from the cell by an external circuit is 0.5 A, what is the power supplied by the cell to the external circuit?

$V = 2\ \text{V},\ I = 0.5\ \text{A}$
$P = IV$
$\quad = 0.5 \times 2$
$\quad = 1\ \text{W}$

Example 28

A cable has a resistance of 0.02 Ω per metre. What is the power loss when a current of 100 mA passes through 10 metres of the cable?

$R = 0.02\ \Omega$ per metre $= 0.02 \times 10 = 0.2\ \Omega$
$I = 100\ \text{mA} = 0.1\ \text{A}$
$P = I^2R$
$\quad = 0.1^2 \times 0.2$
$\quad = 0.002\ \text{W}\ (2\ \text{mW})$

Example 29

A d.c. source of negligible internal resistance has an e.m.f. of 4.2 V and supplies a current to a circuit consisting of a 50 Ω resistor in parallel with 20 Ω resistor. What is the power dissipated in (a) the 50 Ω resistor and (b) the total circuit?

(a) The potential difference across the 50 Ω resistor is 4.2 V. Hence:

$$P = V^2/R$$
$$= 4.2^2/50$$
$$= 0.353 \text{ W}$$

(b) The potential difference across the 20 Ω resistor is 4.2 V. Hence:

$$P = V^2/R$$
$$= 4.2^2/20$$
$$= 0.882 \text{ W}$$

Hence the total power is $0.353 + 0.882 = 1.235$ W.

We could have determined this answer by calculating the equivalent resistance of the parallel arrangement and then used $P = V^2/R$.

$$\frac{1}{R} = \frac{1}{50} + \frac{1}{20}$$

$$\frac{1}{R} = \frac{2 + 5}{100}$$

$$R = \frac{100}{7} \ \Omega$$

$$P = \frac{4.2^2}{100/7}$$

$$= 1.235 \text{ W}$$

Example 30

A current of 200 mA is taken by a circuit consisting of a 10 Ω resistor in series with a parallel combination of a 20 Ω and a 40 Ω resistor. What is the power dissipated in (a) each resistor and (b) the total circuit?

(a) Figure 35 shows the circuit. For the 10 Ω resistor $I = 200$ mA $= 0.20$ A and $R = 10$ Ω, hence:

$$P = I^2R$$
$$= 0.2^2 \times 10$$
$$= 0.4 \text{ W}$$

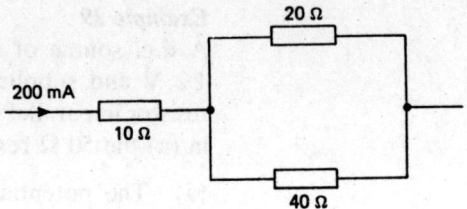

Figure 35

For the parallel arrangement $R_1 = 20\ \Omega$, $R_2 = 40\ \Omega$

$$\frac{1}{R} = \frac{1}{R_1} + \frac{1}{R_2}$$

$$\frac{1}{R} = \frac{1}{20} + \frac{1}{10}$$

$$\frac{1}{R} = \frac{3}{20}$$

$$R = \frac{20}{3}\ \Omega$$

Hence the potential difference across the equivalent resistor R must be:

$$V = 0.2 \times \frac{20}{3}$$

$$= 1.33\ \text{V}$$

The power for the 20 Ω resistor is thus given by:

$$P = \frac{V^2}{R}$$

$$= \frac{1.33^2}{20}$$

$$= 0.088\ \text{W}$$

The power for the 40 Ω resistor is given by:

$$P = \frac{V^2}{R}$$

$$= \frac{1.33^2}{40}$$

$$= 0.044\ \text{W}$$

(b)   The total power for the circuit is thus
0.4 + 0.088 + 0.044 = 0.532 W.

Self-assessment questions

36   What is the power taken from a d.c. source when it drives a current of 500 mA through a resistance of 100 Ω?

37   A potential difference of 15 V is applied across a resistor of 50 Ω. What is the power developed in the resistor?

38   A potential difference of 4 V is applied to a circuit consisting of two resistors in series. If these have resistances of 20 Ω and 50 Ω, what is the power developed in (a) each resistor and (b) the total circuit?

39   A circuit consists of two resistors in parallel. If these have resistances of 20 Ω and 10 Ω and the current taken by the circuit is 0.5 A, what is the power developed in (a) each resistor and (b) the total circuit?

40   A circuit consists of a 100 Ω resistor in series with a parallel combination of a 40 Ω and a 100 Ω resistor. If the current taken by the entire circuit is 150 mA what is the power developed by (a) each resistor and (b) the total circuit?

---

2 Electric fields and capacitance

After reading this chapter you should be able to:

1 Recognize that charged objects exert forces on each other.
2 Define current as rate of movement of charge and calculate charge moved in a circuit.
3 Define electric field strength, electric potential and electrical potential difference, and relate these to the concept of an electric field.
4 Derive and use the relationship between electric field strength and potential gradient.
5 Explain what is happening when a capacitor is charged and then discharged.
6 Define capacitance and solve problems involving capacitors.
7 Describe the make-up and characteristics of practical capacitors.
8 Given details regarding a parallel plate capacitor, calculate its capacitance.
9 Describe multiplate and variable capacitors.
10 Explain the term dielectric strength and relate it to the working voltage of a capacitor.
11 Derive and use the relationship for the energy stored by a capacitor.
12 Solve problems involving capacitors in series and in parallel.

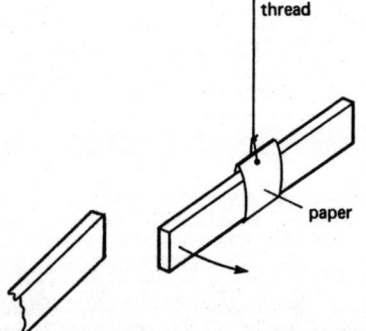

Figure 36 *Repulsion between two like charged objects*

Charge

If you rub a strip of polythene with a cloth and hold the strip near small pieces of paper then the small pieces of paper are attracted towards the strip. We say that the polythene has become charged. You can try a similar experiment with a pen or indeed any plastic strip of material.

If two strips of polythene are rubbed with the same cloth, each strip of polythene becomes charged. If one of these strips is suspended, as shown in Figure 36, and the other charged strip of polythene brought near to it, then the strips will be found to repel

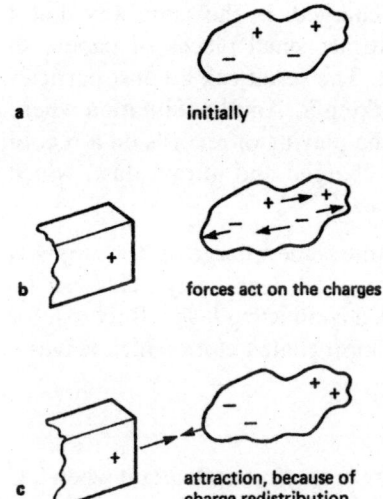

a   initially

b   forces act on the charges

c   attraction, because of
                     charge redistribution

Figure 37

each other. Both the polythene strips were charged in the same way and so have the same type of charge. The conclusion is that *like charged objects repel each other*.

If a strip of cellulose acetate is rubbed by a piece of cloth, it also becomes charged. If this charged strip is then brought near to the suspended charged strip of polythene then the strips are found to attract each other. We thus consider that the cellulose acetate strip carries a different type of charge. *Unlike charged objects attract each other*.

The two types of charge are called positive and negative and the force that occurs between charged objects is called an electric force.

The attraction that occurs between small pieces of paper and a charged strip of material can be explained (see Figure 37) by considering that the paper contains both positive and negative charges in equal amounts. Initially these charges are all mixed up in the paper. However, when a charged strip is brought near to the paper, those charges which are the same as that on the charged strip are repelled and those which are unlike are attracted. The result is that the edge of the paper nearest to the charged strip has more unlike charges and that furthest from the strip more like charges. Thus the nearest edge of the paper is attracted to the charged strip.

Polythene, cellulose acetate, plastics in general and paper are examples of what are called insulators. Such materials will hold a charge. If you charge one end of a strip of polythene, that end can retain its charge for quite a long period of time, provided you do not touch the end. You are a reasonably good conductor of charge and offer a ready path for the charge to flow away from the strip when you touch the end that was charged. Materials through which charge moves very readily are called conductors. Metals are examples of good conductors.

The rate at which charge flows past a point is called current. It does not matter whether the charge is produced by rubbing two objects together or by chemical action in a cell.

> current = rate of movement of charge

The unit used for current is the ampere and that used for charge is the coulomb (C). A rate of charge movement of one coulomb per second is a current of one ampere.

This charging of objects by rubbing one against another can cause problems in many industrial processes. If a process, for example, involves a material passing over a roller, then the contact may result in the material becoming charged. In the same way that a charged strip of material can attract small pieces of paper, so charged material can attract dust. The result can be dust particles adhering to the material and marking it. Another situation where charges can be a problem, is in the playing of records on a record player. The record can become charged and attract dust, which can affect the sound reproduction.

One way of getting rid of troublesome charge is to supply a conducting path so that the charge can leak away. This can be done by coating the material with a conducting layer. Records, for example, can be wiped with an impregnated cloth which leaves a conducting layer on the record.

Example 1
What is the current in a conductor in an electrical circuit when 2 C passes through it every second?

current = rate of movement of charge
$$= 2 \text{ C/s}$$
$$= 2 \text{ A}$$

Example 2
When a current of 200 mA flows through a wire, how much charge is passing through that wire every second?

current = rate of movement of charge
$$= 200 \text{ mA}$$
$$= 200 \times 10^{-3} \text{ A}$$
$$= 200 \times 10^{-3} \text{ C/s}$$
$$= 200 \text{ mC/s}$$

Example 3
A steady current of 0.4 A flows through a wire for 1 minute. How much charge has flowed through that wire in that time?

current = rate of movement of charge
$$= 0.4 \text{ A}$$
$$= 0.4 \text{ C/s}$$

In 1 minute (60 s) the charge flow is $0.4 \times 60 = 24$ C.

Self-assessment questions

1   When two charged objects come close together they repel each other, what can be said about the type of charge carried by the two objects?

2   Give an example of (a) an insulator and (b) a conductor.

3 When a current of 4 A flows through a conductor in an electrical circuit, how much charge is passing through the conductor per second?

4 Over a period of 40 s a charge of 100 mC is found to pass a point in a circuit. What is the current?

5 A steady current of 400 mA flows through a wire in a circuit for a time of 2 minutes. What is the charge that has flowed through the wire in that time?

Electric fields

A charged strip of material can exert a force on another charged strip when they are some distance apart. This is referred to as an electric force. If we took a charged object and put it at some point where it was acted on by a force because it was charged we could say it was acted on by an electric force, perhaps caused by another charged object nearby. Alternatively we could say that the charged object must be in an electric field. By definition an electric field is a region where a charged object experiences an electric force.

Electric field strength (E) is defined as the force per unit charge at a point.

$$E = \frac{F}{Q}$$

where Q is the charge placed at the point and F the electric force experienced by it. If the charge is in coulombs and the force in newtons, the electric field strength is in newtons per coulomb (N/C). There is another equivalent unit which we will meet later in this chapter.

The direction of the electric field at a point is the direction of the force that would be experienced by a positively charged object placed at the point.

An electric field can be represented pictorially by electric lines of force. Such lines map the directions of the electric field, rather like magnetic lines of force map the directions of the magnetic field. Figure 38 shows the lines of force round charged spheres.

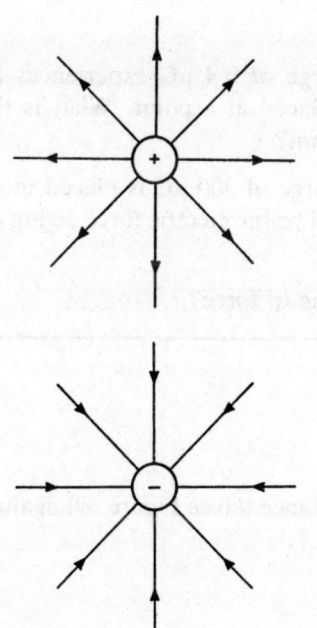

Figure 38 *Lines of force near a charged spherical object*

Example 4
A small object carrying a charge of 2 μC experiences an electric force of 5 N when placed at a particular point. What is the electric field strength at that point?

$$F = 5 \text{ N}, \ Q = 2 \ \mu\text{C} = 2 \times 10^{-6} \text{ C}$$

$$E = \frac{F}{Q}$$

$$= \frac{5}{2 \times 10^{-6}} \qquad\qquad \frac{\text{N}}{\text{C}}$$

$$= 2.5 \times 10^6 \text{ N/C}$$

Example 5

A small object carrying a charge of 20 mC is placed in an electric field of strength 40 N/C. What is the electric force acting on the object?

$$Q = 20 \text{ mC} = 20 \times 10^{-3} \text{ C}, \ E = 40 \text{ N/C}$$

$$E = \frac{F}{Q}$$

$$F = EQ$$
$$= 40 \times 20 \times 10^{-3} \qquad\qquad (\text{N/C}) \times \text{C}$$
$$= 0.80 \text{ N}$$

Self-assessment questions

6 Define electric field strength.

7 A small object carrying a charge of 0.4 µC experiences an electric force of 0.1 N when placed at a point. What is the electric field strength at that point?

8 If a small object carrying a charge of 300 µC is placed in an electric field of 20 N/C, what will be the electric force acting on it?

9 What is shown by an electric line of force?

Electric potential

If you move an object through a distance d (see Figure 39) against a force F then work is done.

$$\text{work} = \text{force} \times \text{distance}$$
$$= F \times d$$

But an object carrying a charge Q in an electric field of strength E experiences a force F, where:

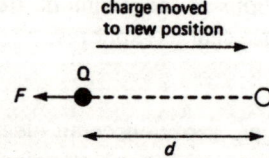

Figure 39

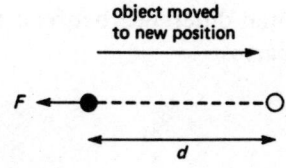

Figure 40

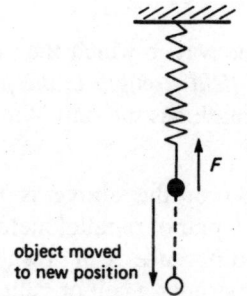

Figure 41

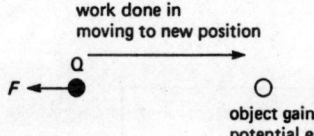

Figure 42

$$E = \frac{F}{Q}$$

and so $F = EQ$

Hence moving a charged object against the electric force in an electric field (see Figure 40) involves work.

Thus work $= (EQ) \times d$

If you have an object suspended from the end of a spring (see Figure 41) and you change the position of the object by pulling the object down to a new position, in doing so stretching the spring, then work is done. You are putting energy into the system. The object in its new position is said to have potential energy, which is equal to the energy you put in to the system. Potential energy is the term given to describe the energy an object has by virtue of its position. If you doubt that the object in its new position has energy, consider what happens when you let go – the object promptly starts to move.

When a charged object is moved in an electric field against the electric force, work is done, just like the case above when an object suspended from a spring was pulled to a new position against the spring force. We can therefore talk of the charged object gaining electric potential energy (see Figure 42). Hence:

change in electric potential energy $= EQd$

The electric potential at a point is defined as the electric potential energy per unit charge at that point.

$$\boxed{\text{electric potential } V = \frac{\text{potential energy}}{Q}}$$

We say there is a potential difference between two points in an electric field when there is a difference in potential energy between the two points.

$$\boxed{\text{potential difference} = \frac{\text{change in potential energy}}{Q}}$$

Potential energy has the unit of joule (J), charge the unit of coulomb (C) and potential or potential difference the unit of volt (V).

For the situation considered above, when a charged object Q is moved against a constant electric force (i.e. the electric field strength is constant), there is a change in potential energy of EQd

in moving a distance d. So the potential difference between the two points in the field a distance d apart is:

potential difference $V = \dfrac{EQd}{Q}$

$$V = Ed$$

or

$$\boxed{E = \dfrac{V}{d}}$$

V/d is the 'potential gradient', i.e. the way in which the potential varies with distance. So the *electric field strength is the potential gradient*. Therefore electric field strength has the unit V/m as well as N/C.

A particularly important application of the above is when a potential difference is applied across a pair of parallel metal plates (see Figure 43). The potential difference can result from connecting a source of d.c. voltage, perhaps a cell or cells, across the plates. If the potential difference between the plates is V and they are a distance d apart, there is a potential gradient of V/d. So between the plates there is an electric field of $E = V/d$. The electric field between the plates is uniform because the potential gradient is uniform.

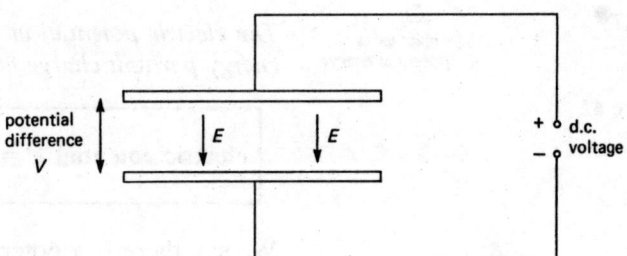

Figure 43

This section on potential, dealing with charges moving in electric fields, may seem to be rather abstract. However, we can apply the ideas to charges moving in wires and through resistors, i.e. electric currents in circuits. When a voltmeter is placed across a resistor and measures the potential difference, what it is determining is the change in potential energy for a charge moving through the resistor, in other words, the work done to move the charge through the resistor.

The potential difference across a component is the work done in moving unit charge through that component.

A cell is a source of energy for a circuit.

The e.m.f. of a cell is the energy converted into electrical energy when unit charge passes through it.

This energy can come from a chemical reaction.

Kirchhoff's second law (that around any closed loop the algebraic sum of the e.m.f.s and the potential differences is zero) is, therefore, simply stating that the energy converted into electrical energy per unit charge in the sources of e.m.f. is equal to the sum of the work done in moving the unit charge through the closed loop. In other words, energy is conserved.

Example 6

A potential difference of 1 kV is applied between two parallel plates a distance of 5 mm apart. What is the electric field between the plates?

$V = 1$ kV $= 1000$ V, $d = 5$ mm $= 5 \times 10^{-3}$ m

$$E = \frac{V}{d}$$

$$= \frac{1000}{5 \times 10^{-3}} \qquad\qquad \frac{V}{m}$$

$$= 200\,000 \text{ V/m (200 kV/m)}$$
$$= 200\,000 \text{ N/C (200 kN/C)}$$

Example 7

What is the work done in moving a charge of 50 μC through a potential difference of 100 V?

$Q = 50$ μC $= 50 \times 10^{-6}$ C, $V = 100$ V

$$\text{potential difference} = \frac{\text{change in potential energy}}{Q}$$

$$V = \frac{\text{work done}}{Q}$$

Hence work done $= VQ$
$$= 100 \times 50 \times 10^{-6} \qquad\qquad V \times C$$
$$= 5 \times 10^{-3} \text{ J}$$

Self-assessment questions

10 Define (a) electric potential and (b) electric potential difference.

11   Explain what is meant by a potential gradient and how it is related to the electric field strength.

12   A potential difference of 20 V is applied between two parallel plates a distance of 2 mm apart. What is the electric field between the plates?

13   What is the work done in moving a charge Q through a potential difference of V?

14   What is the work done in moving 20 μC through a potential difference of 12 V?

15   A voltmeter connected across a resistor gives a reading of 2 V. What is the work done in moving 0.5 C of charge through that resistor?

Capacitance

Figure 44 shows a circuit in which one of the components is a pair of parallel plates. The gap between them is an insulator such as air. This parallel plate arrangement is known as a capacitor. When the switch is closed the meters on both sides of the capacitor register a current. The current is initially high but decays with time to eventually become zero.

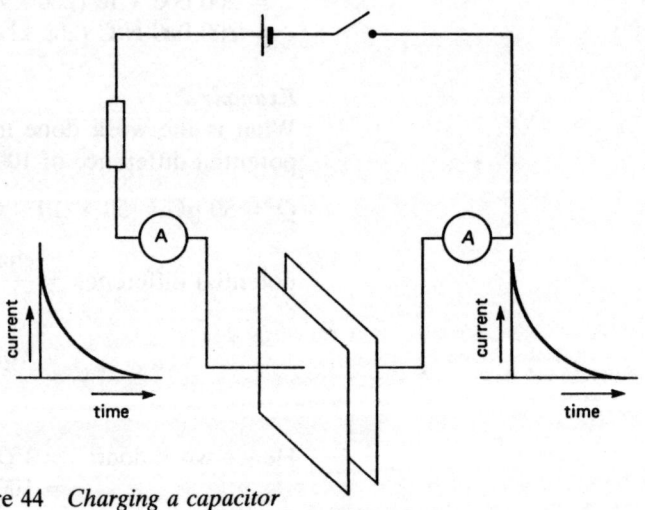

Figure 44   *Charging a capacitor*

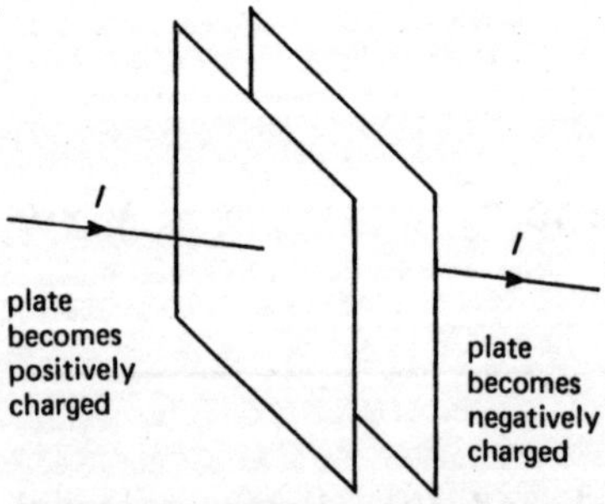

For one of the plates the current direction is towards the plate, for the other plate the current direction is away from the plate (see Figure 45). When a current exists there is a movement of charge, so one plate has charge flowing into it and the other plate has

plate becomes positively charged

plate becomes negatively charged

Figure 45   *Charging a capacitor*

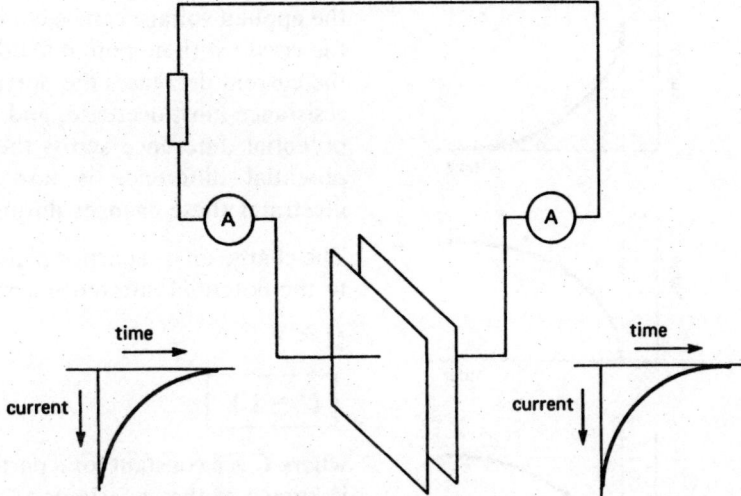

Figure 46 *Discharging a capacitor*

charge flowing away from it. One plate therefore becomes positively charged and the other plate negatively charged. The capacitor in this state is said to be charged.

If the charged capacitor is then connected, as shown in Figure 46, the meters again record a current. The current is initially high, but in the opposite direction to the previous current, and decays with time to eventually become zero. We can explain this by considering the capacitor to have discharged: the positive charge on one plate of the charged capacitor cancels out the negative charge on the other plate when the two are connected together.

If a current of 2 A flows for 1 s, the charge movement is 2 C, current being rate of movement of charge. By considering how the current varies with time when the capacitor is charged the amount of charge moved on to one plate of the capacitor and off the other plate can be estimated. The size of the charge moved on to one plate is the same as the size of the charge moved off the other plate. Their only difference is that one is positive and the other negative.

The larger the voltage used to charge the capacitor the greater the charge on the fully charged capacitor plates. Doubling the voltage doubles the charge. The charge is proportional to the voltage.

$Q \propto V$

When the current first starts to flow in the charging operation the capacitor has no charge on its plates. In the course of time, the current decreases as the charge builds up on the plates. Finally there is no current, but maximum charge on the plates. Initially all

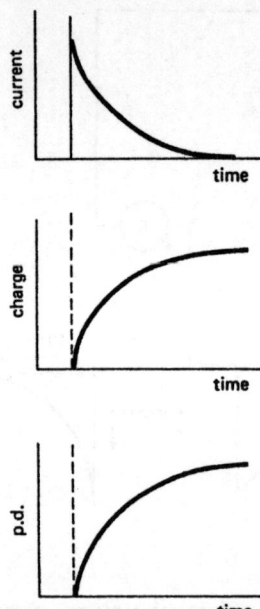

Figure 47   *Charging a capacitor*

the applied voltage can be considered to be across the resistance in the circuit with no potential difference across the capacitor, but as the current decreases the potential difference ($V = IR$) across the resistance must decrease, and when the current becomes zero the potential difference across the resistor is zero. However, all the potential difference is now across the capacitor. Figure 47 illustrates these changes during charging.

The charge on a capacitor plate is, therefore, directly proportional to the potential difference across the capacitor:

$$Q \propto V$$

$$\boxed{Q = CV}$$

where C is a constant for a particular parallel plate arrangement. C is known as the capacitance.

When the charge is in coulombs (C) and the potential difference in volts (V) then the capacitance is in farads (F). A farad is a very large unit of capacitance and most capacitors are microfarads or less.

Not only parallel plate arrangements have capacitance: any isolated conductor has capacitance. Thus the parallel plates in the circuit referred to above could be replaced by two spheres, provided they are isolated from each other and not in electrical contact with each other. We could have had a sphere for one and a plate for the other or a sphere for one and the ground for the other.

Example 8
What is the charge on the plates of a 8 μF capacitor when it is fully charged by a voltage of 20 V?

$C = 8\ \mu\text{F} = 8 \times 10^{-6}\ \text{F},\ V = 20\ \text{V}$
$Q = CV$
   $= 8 \times 10^{-6} \times 20$ F × V
   $= 1.6 \times 10^{-4}\ \text{C}$

On one plate there will be a charge of $+ 1.6 \times 10^{-4}$ C and on the other plate a charge of $- 1.6 \times 10^{-4}$ C.

Self-assessment questions
16 Define capacitance.

17 What is the unit of capacitance?

18 What is the charge on the plates of a 16 μF capacitor when it is fully charged by a voltage of 12 V?

19 A 500 µF capacitor that has been fully charged by a voltage of 12 V is discharged. What is the amount of charge that will flow off each plate during the total discharge?

Capacitance of a parallel plate capacitor

There are many forms of what are essentially parallel plate capacitors. Figure 48 illustrates some of them. In the paper capacitor the parallel plates are sheets of aluminium foil which are kept apart and insulated from each other by sheets of waxed paper. With the ceramic capacitors, silver plates have a ceramic disc to keep them apart and insulated from each other. In the electrolytic capacitor there are two electrodes in an electrolyte. When a current is first passed through the arrangement a thin oxide layer is formed on one of the electrodes. As a result, the two conductors are separated by a thin layer of oxide, which acts as an insulator. With a paper or a ceramic capacitor it does not matter which of the plates is connected to the positive side of the supply and which to the negative side. With an electrolytic capacitor it does matter: they can only be used when connected the correct way round in a circuit.

Paper capacitors tend to have capacitances in the range of 50 pF (10^{-12} F) to a few microfarads. Ceramic capacitors have values in the region of 1000 pF to 0.5 µF. Electrolytic capacitors range from about 1 µF to many thousands of microfarads.

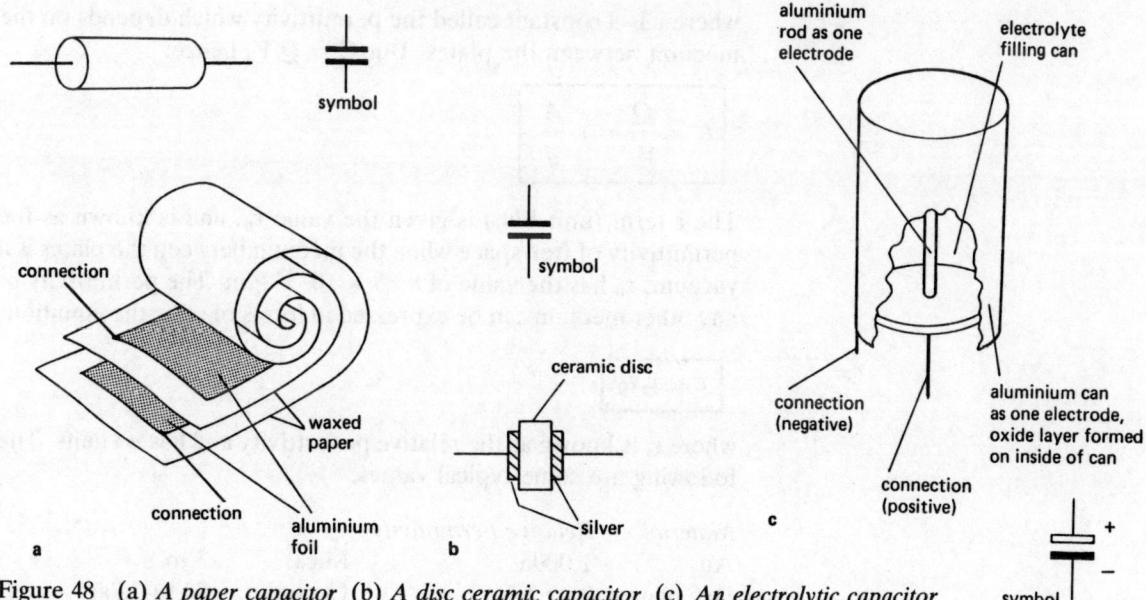

Figure 48 (a) *A paper capacitor* (b) *A disc ceramic capacitor* (c) *An electrolytic capacitor*

What factors determine the capacitance of a parallel plate capacitor? Experimentally it can be found that:

1   The charge on a plate is directly proportional to the potential difference V between the plates

$$Q \propto V$$

2   If the potential difference is kept constant and the plate separation varied then it is found that the charge on a plate is inversely proportional to the plate separation d

$$Q \propto \frac{1}{d}$$

3   If both the potential difference and the plate separation are kept constant, but the area of the plates changed, then it is found that the charge on a plate is directly proportional to the area

$$Q \propto A$$

4   If the potential difference, plate separation and plate area are kept constant, but the medium between the plates is changed, then the charge on the plates is found to change.

All the above factors can be combined together in one equation:

$$Q = \varepsilon \frac{VA}{d}$$

where ε is a constant called the permittivity which depends on the medium between the plates. But $C = Q/V$, hence:

$$\boxed{C = \frac{Q}{V} = \varepsilon \frac{A}{d}}$$

The ε term (unit F/m) is given the value ε_0, and is known as the permittivity of free space when the medium between the plates is a vacuum. ε_0 has the value of 8.85×10^{-12} F/m. The permittivity of any other medium can be expressed in terms of ε_0 by the equation:

$$\boxed{\varepsilon = \varepsilon_r \varepsilon_0}$$

where ε_r is known as the relative permittivity and has no units. The following are some typical values:

Material	Relative permittivity		
Air	1.0006	Mica	3 to 8
Dry paper	2 to 2.5	Ceramics	80 to 1200

The term dielectric is often used for the material inserted between the parallel plates in a capacitor.

Example 9

Calculate the capacitance of a capacitor which consists of two parallel plates, each of area 100 cm², when the space between them is filled with a material of thickness 0.4 mm and having a relative permittivity of 2.

$A = 100$ cm² $= 100 \times 10^{-4}$ m², $d = 0.4$ mm $= 0.4 \times 10^{-3}$ m, $\varepsilon_r = 2$

$$C = \varepsilon_r \varepsilon_0 \frac{A}{d}$$

The value of ε_0 is 8.85×10^{-12} F/m. Hence:

$$C = 2 \times 8.85 \times 10^{-12} \times \frac{100 \times 10^{-4}}{0.4 \times 10^{-3}} \qquad \text{(F/m)} \times \frac{\text{m}^2}{\text{m}}$$

$$= 4.4 \times 10^{-10} \text{ F (44 nF or 4400 pF)}$$

Example 10

Calculate the capacitance of a capacitor which consists of two parallel plates, each of area 0.01 m² spaced 0.2 mm apart when the space between them is just occupied by air.

$A = 0.01$ m², $d = 0.2$ mm $= 0.02 \times 10^{-3}$ m, ε_r can be taken as 1 in the absence of more specific information.

$$C = \varepsilon_r \varepsilon_0 \frac{A}{d}$$

$$= 1 \times 8.85 \times 10^{-12} \times \frac{0.01}{0.02} \times 10^{-3} \qquad \text{(F/m)} \times \frac{\text{m}^2}{\text{m}}$$

$$= 4.4 \times 10^{-9} \text{ F (4.4 nF or 4400 pF)}$$

Self-assessment questions

20 What factors determine the capacitance of a parallel plate capacitor?

21 What information would you need for connecting an electrolytic capacitor in a circuit that would not be needed with a paper or ceramic capacitor?

22 Calculate the capacitance of a capacitor which consists of two parallel plates, each of area 0.02 m² spaced 0.5 mm apart, when the space between them is filled with (a) air and (b) a dielectric of relative permittivity 7.

23 A capacitor consists of two parallel metal plates, each of area 60 cm², and separated by a dielectric of thickness 0.1 mm and having a relative permittivity of 3.2. Calculate its capacitance.

24 What area plates are needed for a parallel plate ceramic capacitor if it is to have a capacitance of 0.1 nF and the dielectric has a thickness of 1 mm and a relative permittivity of 1000?

Multiplate and variable capacitors

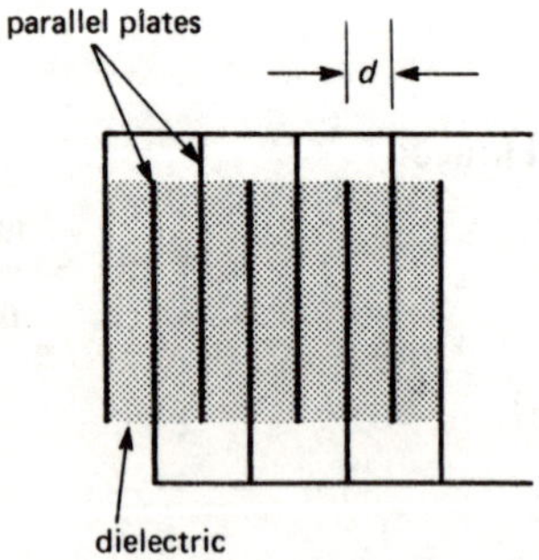

In order to obtain a high capacitance, the area of the plates needs to be large. This can lead to inconveniently large objects. An alternative to having two large area plates for the capacitor, is to use a number of sets of parallel plates. Figure 49 shows the type of arrangement used. Each plate is separated from the next one by a layer of dielectric. In the figure there are 8 plates and if you look at how many pairs of parallel plates that arrangement gives you will find there to be 7. If there are n plates there are $(n - 1)$ pairs of parallel plates. The effective plate area with n plates thus becomes $(n - 1)A$, with A being the area of one plate, and so the capacitance is

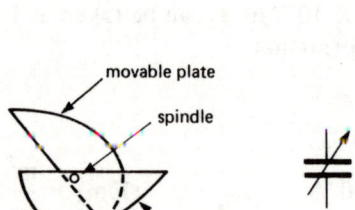

parallel plates

d

dielectric

Figure 49 *A multiplate capacitor*

$$C = \varepsilon_r \varepsilon_0 \frac{(n - 1)A}{d}$$

In some circuit applications there is a need for a variable capacitor. One method adopted to give a variable capacitor is to vary the plate area. Figure 50 shows how this can be done. The plates are mounted on a spindle so that one plate can be rotated with respect to the other and so vary the area of plate directly opposite another plate.

movable plate

spindle

fixed plate

Figure 50 *A variable capacitor*

Example 11
A multiplate capacitor has 11 plates, each with an area of 50 cm² and separated by a dielectric of thickness 0.4 mm and relative permittivity 3. What is the capacitance?

$A = 50$ cm² $= 50 \times 10^{-4}$ m², $d = 0.4$ mm $= 0.4 \times 10^{-3}$ m, $\varepsilon_r = 3$

$$C = \varepsilon_r \varepsilon_0 \frac{(n - 1)A}{d}$$

$$= \frac{3 \times 8.85 \times 10^{-12} \times (11 - 1) \times 50 \times 10^{-4}}{0.4 \times 10^{-3}} \quad (\text{F/m}) \times \frac{\text{m}^2}{\text{m}}$$

$$= 3.3 \times 10^{-9} \text{ F}$$

Self-assessment questions

25 A multiplate capacitor has 9 plates, each with an area of 100 cm^2 and separated by a dielectric of thickness 0.2 mm and relative permittivity 4. What is the capacitance?

26 A multiplate capacitor is to have a capacitance of 8.85 nF. How many plates of area 250 cm^2 are needed if the dielectric which separates the plates is to have a thickness of 0.5 mm and a relative permittivity of 2.5?

Dielectric strength

When a potential difference V is maintained across the dielectric in a parallel plate capacitor we have in fact produced an electric field in the dielectric, since there is a potential gradient.

$$\text{electric field strength} = \frac{V}{d}$$

where d is the thickness of the dielectric, i.e. the separation between the two parallel plates. For example, if the potential difference applied to the capacitor is 6 V and the thickness of the dielectric is 0.2 mm $(0.2 \times 10^{-3} \text{ m})$, then the electric field strength is:

$$E = \frac{6}{0.2 \times 10^{-3}} \quad \frac{\text{V}}{\text{m}}$$

$$= 3 \times 10^4 \text{ V/m}$$

An electric field is however a region where forces are experienced by charged objects (remember $E = F/Q$). The dielectric is composed of atoms which themselves are composed of particles, some of which are charged. The electric field in the dielectric must mean that forces are exerted on the charged particles in the dielectric. Dielectrics are electrical insulators and thus do not normally have charged particles able to move about and give a current. If, however, the electric field is high enough, it is possible for it to pull charged particles out of the dielectric atoms and give a current, i.e. a flow of charge. When this happens the insulation breaks down and the capacitor ceases to function as a capacitor. Generally the dielectric, and hence the capacitor, is ruined.

The dielectric strength is defined as the potential gradient (i.e. the electric field strength) at which the insulation breaks down.

Dielectric strengths are generally quoted in units of kV/mm. The following are some typical values.

Material	Dielectric strength/kV mm^{-1}
Air	3
Dry paper	14
Mica	150

The voltage at which a capacitor will break down is given by:

breakdown voltage = dielectric strength × dielectric thickness

Thus for an air capacitor with a plate separation of 2 mm the breakdown voltage is:

breakdown voltage $= 3 \times 2$ (kV/mm) × mm
$= 6$ kV

Commercial capacitors have a safe working voltage quoted for them. This voltage is below the breakdown voltage to allow a margin for safety.

Example 12
A capacitor has a dielectric of thickness 0.4 mm. What is the electric field in the dielectric when a potential difference of 12 V is applied to the capacitor?

$V = 12$ V, $d = 0.4$ mm $= 0.4 \times 10^{-3}$ m

$$E = \frac{V}{d}$$

$$= \frac{12}{0.4 \times 10^{-3}} \quad \frac{V}{m}$$

$$= 3 \times 10^4 \text{ V/m}$$

Example 13
The dielectric strength of dry paper is 14 kV/mm and its thickness 0.1 mm. What would be the breakdown voltage for a capacitor using this paper as a dielectric?

dielectric strength $= 14$ kV/mm, $d = 0.1$ mm.
breakdown voltage $= 14 \times 0.1$ (kV/mm) × mm
$= 1.4$ kV

Self-assessment questions

27   A potential difference of 100 V is applied between the parallel plates of a capacitor, the plates being 0.2 mm apart. What is the electric field strength between the plates?

28 Define the term dielectric strength.

29 What is the breakdown voltage for a capacitor with a dielectric of thickness 0.2 mm and a dielectric strength of 14 kV/mm?

30 Why do capacitors have 'working voltages' quoted for them?

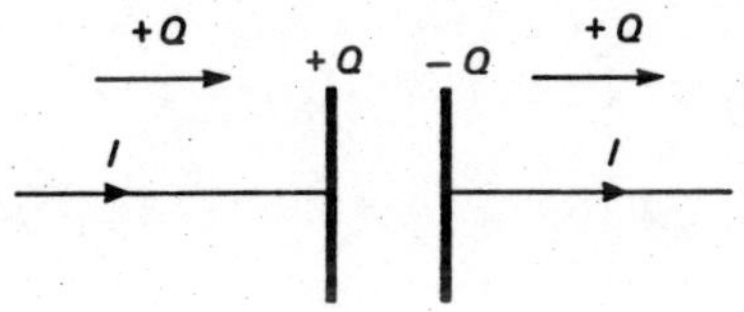

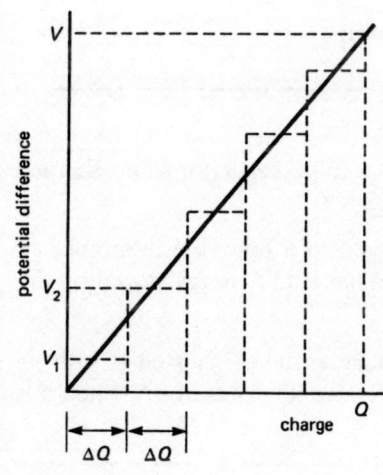

Figure 51 *Charging a capacitor*

Energy stored in a capacitor

Figure 51 shows the situation that occurs during the charging of a capacitor. At some time the capacitor plates have a charge Q and the charging current is pushing more charge along the connecting wires. For the plate that has a positive charge, we can consider the current to be pushing positive charge towards it. Naturally there will be repulsive forces between the like charges. Thus to get more charge on to the plate, work has to be done to overcome the repulsive force. In the case of the other plate, which is negatively charged, positive charge is being removed. Work has to be done in this case to overcome the attractive forces and so remove the charge. Thus to charge a capacitor work has to be done.

When the charge on a capacitor plate increases, so does the potential difference across the capacitor. But (see page 50):

$$\text{potential difference } V = \frac{\text{change in potential energy}}{\text{charge moved } Q}$$

Thus: change in potential energy $= QV$

Because not all the charge Q is being moved on to the capacitor when the potential difference is V, we cannot just use the above equation to arrive at a relationship for the total energy. Figure 52 shows how the potential difference varies with the charge (remember the relationship $Q = CV$: this graph is a plot of V against Q from that relationship). The energy involved in adding a small charge ΔQ when the potential difference is V_1 is $\Delta Q \times V_1$. This is the area of the graph in the first strip on the graph. The energy involved in adding a charge ΔQ when the potential difference has risen to V_2 is $\Delta Q \times V_2$, the area of the second strip on the graph. We can find the energy involved in adding the total charge Q by adding up the small bits of energy owing to each of the small bits of charge ΔQ. This sum is the total area under the graph, i.e. the area of the triangle of base Q and height V. This is an area of $\frac{1}{2}QV$. Thus the energy needed to put a charge Q on to the plates of a capacitor is $\frac{1}{2}QV$.

Figure 52

When the capacitor is allowed to discharge, this energy is put back into the circuit. The energy released is thus $\frac{1}{2}QV$. Hence the charged capacitor stores energy.

energy stored $= \frac{1}{2}QV$

As $Q = CV$ this equation can be written in a number of forms.

$$\text{energy stored} = \tfrac{1}{2}QV = \tfrac{1}{2}CV^2 = \tfrac{1}{2}Q^2/C$$

Example 14

What is the energy stored by an 8 µF capacitor when charged to a potential difference of 20 V?

$C = 8\ \mu\text{F} = 8 \times 10^{-6}\ \text{F},\ V = 20\ \text{V}$
energy stored $= \frac{1}{2}CV^2$
$$= \tfrac{1}{2} \times 8 \times 10^{-6} \times 20^2 \qquad\qquad \text{F} \times \text{V}^2$$
$$= 1.6 \times 10^{-3}\ \text{J}$$

Example 15

A capacitor is required to store an energy of 20 mJ when a voltage of 30 V is applied to it. What capacitance is required?

energy $= 20\ \text{mJ} = 20 \times 10^{-3}\ \text{J},\ V = 30\ \text{V}$
energy stored $= \frac{1}{2}CV^2$

Hence $\qquad C = \dfrac{2 \times \text{energy}}{V^2}$

$$= \dfrac{2 \times 20 \times 10^{-3}}{30^2} \qquad\qquad \dfrac{\text{J}}{\text{V}^2}$$

$$= 4.4 \times 10^{-5}\ \text{F}\ (44\ \mu\text{F})$$

Self-assessment questions

31 What is the energy stored by a 16 µF capacitor when charged to a potential difference of 12 V?

32 A capacitor when fully charged to a potential difference of 20 V has a charge of 2 µC. What is the energy stored by the capacitor?

33 What is the energy released by a 100 µF capacitor with an initial potential difference between its plates of 6 V when it is fully discharged?

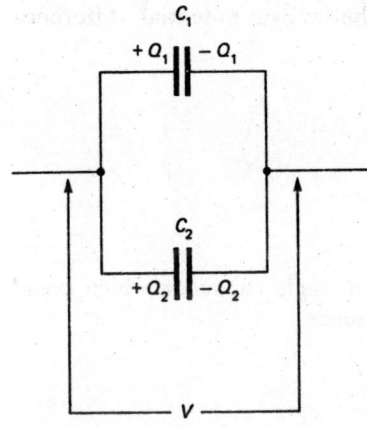

Figure 53 *Capacitors in parallel*

Capacitors in parallel

When capacitors are connected in parallel (see Figure 53), the potential difference across each capacitor will be the same. The charge on capacitor C_1 will be $Q_1 = C_1V$; the charge on capacitor C_2 will be $Q_2 = C_2V$. If the two capacitors are to be replaced by a single capacitor, then the charge on this equivalent capacitor must be $Q_1 + Q_2$.

$$Q = Q_1 + Q_2$$
$$Q = C_1V + C_2V$$
$$\frac{Q}{V} = C_1 + C_2$$

But Q/V is the capacitance C of the equivalent capacitor. Hence:

$$\boxed{C = C_1 + C_2}$$

Example 16
What is the total capacitance when an 8 μF capacitor is connected in parallel with a 16 μF capacitor?

$C_1 = 8$ μF, $C_2 = 16$ μF
$C = C_1 + C_2$
$= 8 + 16$ μF
$= 24$ μF

Self-assessment questions

34 What is the total capacitance when a 20 μF capacitor is connected in parallel with a 10 μF capacitor?

35 Three capacitors are connected in parallel. If they have capacitances of 1 μF, 2 μF and 8 μF, what is the total capacitance?

Capacitors in series

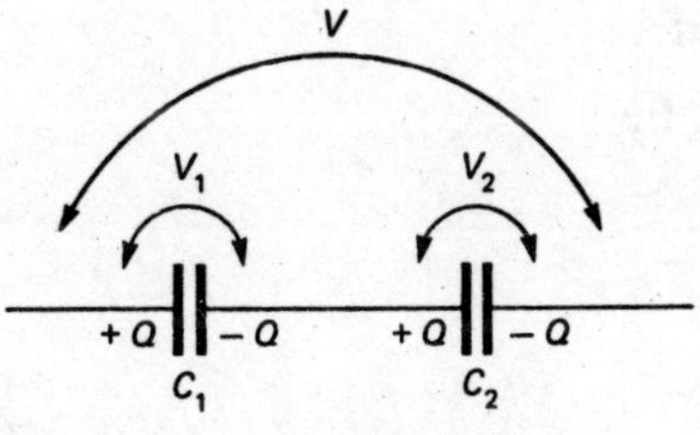

When capacitors are connected in series (see Figure 54) the potential difference across the group of capacitors is equal to the sum of the potential differences across each capacitor.

total potential difference $V = V_1 + V_2$

Because the two capacitors are in series, the same current must flow to C_1 as flows from C_2. Thus the charges on the capacitors must be the same. Hence for C_1 we have $Q = C_1V_1$ and for C_2 we

Figure 54 *Capacitors in series*

have $Q = C_2V_2$. Thus using the above potential difference equation:

$$V = \frac{Q}{C_1} + \frac{Q}{C_2}$$

$$\frac{V}{Q} = \frac{1}{C_1} + \frac{1}{C_2}$$

But Q/V is the capacitance C of a single capacitor which could replace the series arrangement. Hence:

$$\boxed{\frac{1}{C} = \frac{1}{C_1} + \frac{1}{C_2}}$$

Example 17
What is the total capacitance when an 8 µF capacitor is connected in series with a 16 µF capacitor?

$C_1 = 8\ \mu F,\ C_2 = 16\ \mu F$

$$\frac{1}{C} = \frac{1}{C_1} + \frac{1}{C_2}$$

$$\frac{1}{C} = \frac{1}{8} + \frac{1}{16} \qquad\qquad \frac{1}{\mu F}$$

$$\frac{1}{C} = \frac{2+1}{16}$$

$$C = \frac{16}{3}$$

$$C = 5.3\ \mu F$$

Example 18
A potential difference of 20 V is connected across an arrangement of two capacitors in series. If they have capacitances of 2 µF and 8 µF what will be the potential differences across each capacitor?

$V = 20\ V,\ C_1 = 2\ \mu F,\ C_2 = 8\ \mu F$

The total capacitance C is given by:

$$\frac{1}{C} = \frac{1}{C_1} + \frac{1}{C_2}$$

$$\frac{1}{C} = \frac{1}{2} + \frac{1}{8} \qquad\qquad \frac{1}{\mu F}$$

$$\frac{1}{C} = \frac{4 + 1}{8}$$

$$C = \frac{8}{5} \; \mu F$$

$$C = 1.6 \; \mu F$$

Hence the charge on the capacitors must be:

$$Q = CV$$
$$= 1.6 \times 10^{-6} \times 20 \qquad\qquad\qquad F \times V$$
$$= 32 \times 10^{-6} \; C$$

The charge on each capacitor is the same. Hence the potential difference across C_1 is:

$$V_1 = \frac{Q}{C_1}$$

$$= \frac{32 \times 10^{-6}}{2 \times 10^{-6}} \qquad\qquad\qquad \frac{C}{F}$$

$$= 16 \; V$$

The potential difference on the other capacitor must be $20 - 16 = 4$ V. We could however have calculated it in the same way as for C_1, i.e.:

$$V_2 = \frac{Q}{C_2}$$

$$= \frac{32 \times 10^{-6}}{8 \times 10^{-6}} \qquad\qquad\qquad \frac{C}{F}$$

$$= 4 \; V$$

Self-assessment questions

36 What is the capacitance of an arrangement consisting of a 2 μF capacitor in series with a 4 μF capacitor?

37 Three capacitors are connected in series. What is the total capacitance if they have capacitances of 4 μF, 12 μF and 16 μF?

38 If a potential difference of 12 V is connected across an arrangement of a 10 μF capacitor in series with a 20 μF capacitor, what will be the potential difference across each capacitor?

Series-parallel connections

The way to tackle a circuit which involves capacitors in both series and parallel connections is to reduce them to a number of just series and just parallel circuits and solve each one in turn. Figure 55 illustrates how this can be done.

For the first step the equivalent capacitance is found for the capacitances of C_2 and C_3 in series, the result being C_5. The second step involves considering the parallel circuit of C_5 and C_4, the equivalent being determined as C_6. Finally the series circuit of C_6 and C_1 is considered and the equivalent capacitance C_7 found. This then is the capacitance of the entire circuit.

Example 19
Find the capacitance of the circuit given in Figure 56.

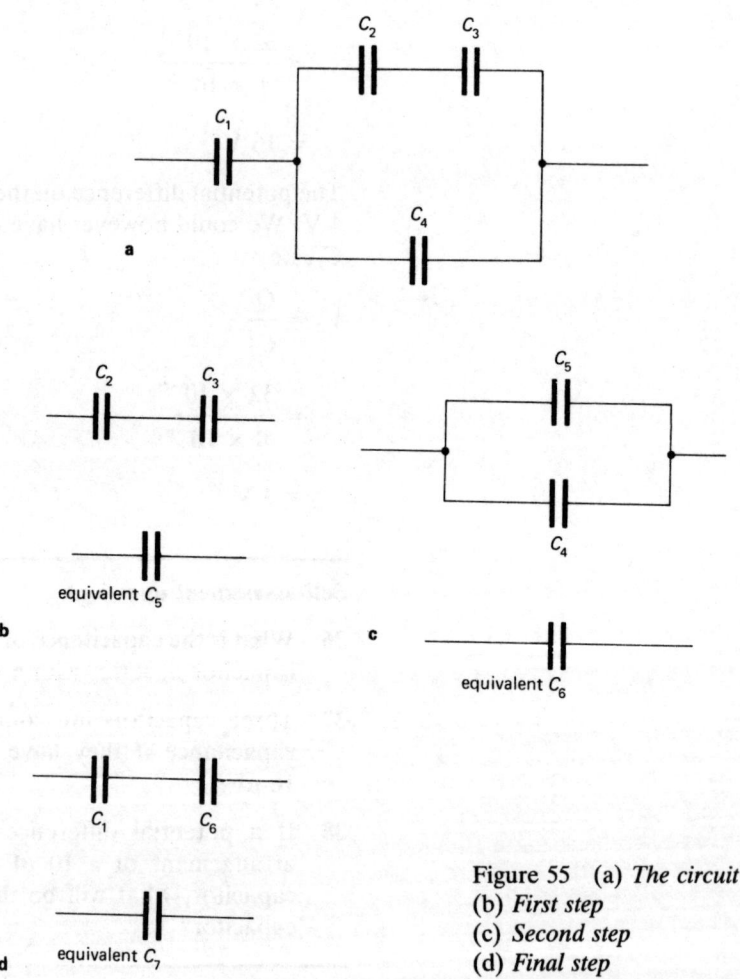

Figure 55   (a) *The circuit*
(b) *First step*
(c) *Second step*
(d) *Final step*

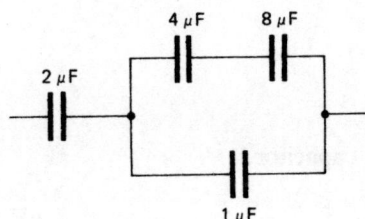

Figure 56

First step: the 4 μF and 8 μF capacitors.

$$\frac{1}{C} = \frac{1}{C_1} + \frac{1}{C_2}$$

$$= \frac{1}{4} + \frac{1}{8} \qquad\qquad \frac{1}{\mu F}$$

$$= \frac{3}{8}$$

$$C = \frac{8}{3} \ \mu F$$
$$= 2.7 \ \mu F$$

Second step: this 2.7 μF capacitance and the 1 μF capacitor.
$$C = C_1 + C_2$$
$$= 2.7 + 1 \qquad\qquad \mu F$$
$$= 3.7 \ \mu F$$

Third step: this 3.7 μF capacitance and the 2 μF capacitor.

$$\frac{1}{C} = \frac{1}{C_1} + \frac{1}{C_2}$$

$$= \frac{1}{3.7} + \frac{1}{2} \qquad\qquad \frac{1}{\mu F}$$

$$= \frac{2 + 3.7}{3.7 \times 2}$$

$$C = \frac{3.7 \times 2}{5.7} \qquad\qquad \mu F$$

$$= 1.3 \ \mu F$$

Example 20
Find the capacitance of the circuit given in Figure 57.

First step: the 1 μF capacitors.
$$C = C_1 + C_2$$
$$= 1 + 1 \qquad\qquad \mu F$$
$$= 2 \ \mu F$$

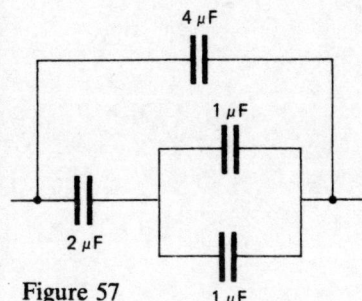

Figure 57

Second step: this 2 μF and the 2 μF in the circuit.

$$\frac{1}{C} = \frac{1}{C_1} + \frac{1}{C_2}$$

$$= \frac{1}{2} + \frac{1}{2} \qquad\qquad \frac{1}{\mu F}$$

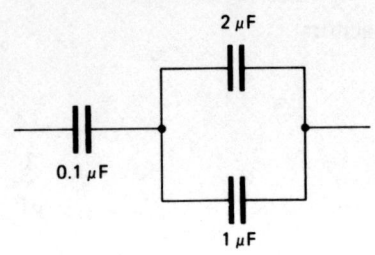

$$= \frac{2}{2}$$

$C = 1 \ \mu F$

Third step: this 1 μF and the 4 μF capacitor.

$C = C_1 + C_2$
$\quad = 1 + 4$ μF
$\quad = 5 \ \mu F$

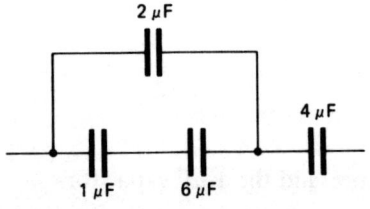

Figure 58

Self-assessment questions

39 Find the capacitances of the circuits shown in Figure 58.

40 If for the circuit in Figure 58 (a) a potential difference of 50 V is connected across the entire arrangement of capacitors, what will be the potential difference across the 0.1 μF capacitor?

3 Basic electromagnetism

After reading this chapter you should be able to:

1 Recognize that magnetic forces exist and that like poles repel, unlike poles attract.
2 Describe the magnetic field pattern for a current-carrying single wire, a loop of wire and a solenoid.
3 Explain and use the equation for the force on a current-carrying conductor, $F = BIL$.
4 Use Fleming's left-hand rule.
5 Explain the forces on a coil in a uniform magnetic field and how the effect is utilized in the moving coil meter and the simple d.c. motor.
6 Explain the terms flux and flux density.
7 Explain the basic principles of electromagnetic induction, including Faraday's laws, Lenz's law and the Neumann equation $E = -d\Phi/dt$.
8 Calculate the average e.m.f. induced when a flux change occurs.
9 Derive and use the equation $E = BLv$, for electromagnetic induction.
10 Explain the operation of the simple a.c. generator.

Revision of basic principles

Magnetic fields can be produced by permanent magnets or by passing a current through a conductor. Magnetic fields can be considered to be regions where forces are experienced by objects because of magnetic effects. Thus if iron filings are sprinkled round a bar magnet, a pattern similar to that shown in Figure 59 is obtained. The iron filings are acted on by forces which cause them to align themselves in particular directions. These directions are said to be the directions of the magnetic field.

When a permanent magnet is suspended, or pivoted, so that it is free to move in the horizontal plane it ends up with one end pointing towards the Earth's north pole and the other end towards the south pole. The end pointing towards the north is known as the north-seeking pole of the magnet, or for short, the north pole. The

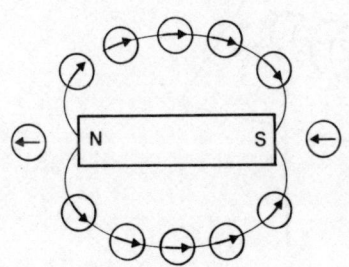

Figure 59 *The magnet field pattern for a bar magnet*

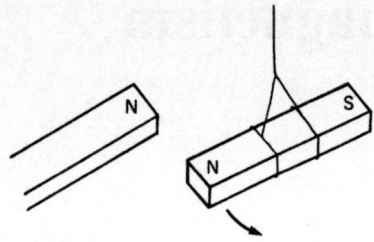

Figure 60   *Repulsion between like poles*

other end is known as the south pole. It is aligning with the Earth's magnetic field direction.

When a magnet is brought close to a suspended or pivoted magnet, repulsion is found to occur (see Figure 60) when like pole is brought close to like pole. Attraction is found to occur when unlike poles are brought close together.

Like poles repel, unlike poles attract.

The direction of the magnetic field at a point is the direction of the magnetic force acting at that point, if a north pole were positioned there. Thus, the arrows on the field pattern in Figure 59 are the field directions, i.e. the directions of the forces on a north pole.

Figure 61 shows the types of magnetic field patterns produced by currents passing through conductors. Figure 61(a) shows a solenoid, a coil which is long in comparison with its diameter. The field pattern for the solenoid is rather similar to that given by a bar magnet.

Figure 61   *Magnetic field patterns for current-carrying conductors*
(a) *A single wire*
(b) *A loop*
(c) *A solenoid*

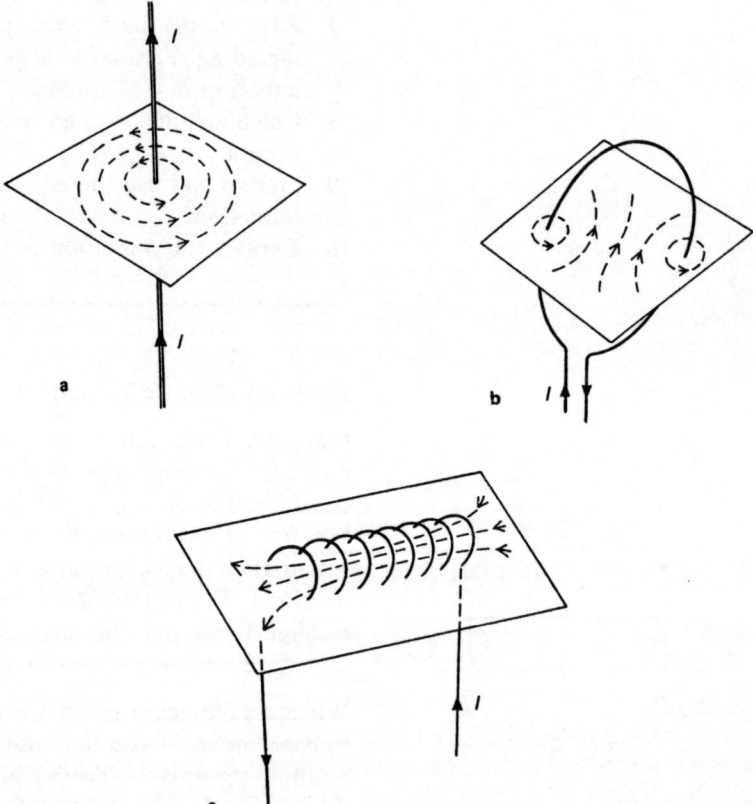

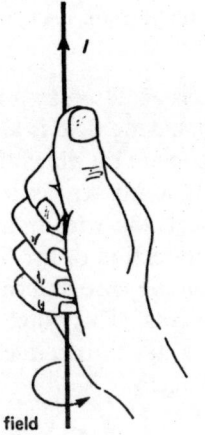

Figure 62

Figure 62 shows a useful way of remembering the direction of the magnetic field near a current-carrying conductor – if you imagine grasping the wire with your right hand, then the thumb which lies along the wire indicates the current direction and the fingers which wrap round the wire indicate the field direction. This can be applied to a current-carrying conductor, whatever its form. However in the case of a solenoid it is often useful, because it behaves so much like a bar magnet, to know which end is acting like a north pole and which like a south pole. Figure 63 shows a useful way of remembering this. If the current direction when the solenoid is seen end on is clockwise, as in (a), then that end of the solenoid is a south pole. If the current direction is anticlockwise, as in (b), then that end of the solenoid behaves as a north pole.

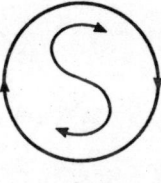
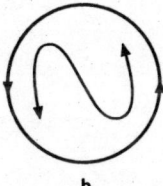

a   b

Figure 63   *End view of solenoid*

Self-assessment questions

1   When one end of a bar magnet is brought close to the north end of a pivoted magnet repulsion occurs. What is the polarity of that end of the bar magnet?

2   For the solenoid shown in Figure 61(c), which end is behaving like a north pole?

3   How would it be possible to determine whether there was a magnetic field present at some place? (No practical details, just the principle.)

Magnetic forces

If, as in Figure 60, a bar magnet is brought near to a suspended magnet then forces act on each magnet. We can say that one magnet produces a magnetic field in the space around it and because the other magnet is in this magnetic field a force acts on it. If we had brought a current-carrying coil near to the suspended magnet, the effect could have been the same. The source of the magnetic field could be a permanent magnet or a current in a conductor.

A bar magnet placed in a magnetic field is acted on by a force. A current-carrying conductor placed in a magnetic field is also acted on by a force. Figure 64 shows an arrangement by which this force can be demonstrated and measured. It is known as a current balance. When the current passes through the wire frame, then part of it is in a magnetic field since the direction of the field is in the horizontal plane. A force then acts on the wire and causes the wire frame to tilt on its knife edge pivots. This force can be determined by measuring the weight that has to be added to the frame to bring it back to the horizontal again.

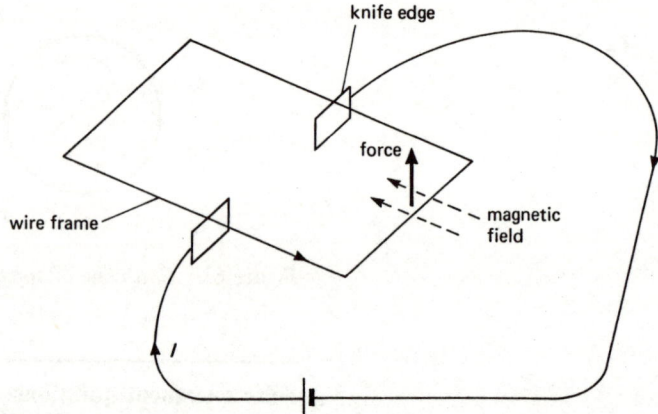

Figure 64   *A current balance*

The force acting on a current-carrying conductor in a magnetic field is found to be directly proportional to the current I in the conductor.

$F \propto I$

If the current is doubled, the force is doubled. The force is also found to be directly proportional to the length of the current-carrying conductor L that is in the magnetic field.

$F \propto L$

If the length is doubled the force is doubled. The force also depends on the strength of the magnetic field. If we denote this 'strength' by the letter B then we can combine the factors into the general equation:

$$F = BIL$$

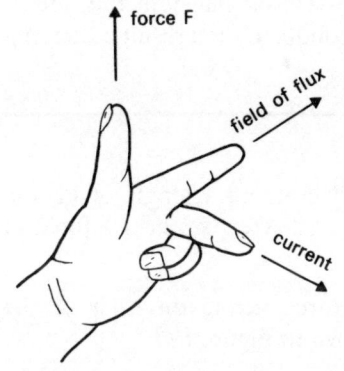

Figure 65 *Fleming's left-hand rule*

The term B is generally known as the flux density, which will be explained on page 74. The above equation indicates $B = F/IL$, so the unit for B could be newton per amp metre (N A^{-1} m^{-1}). A more usual unit is weber per square metre (Wb/m^2) or tesla (T).

$$1 \text{ N A}^{-1} \text{ m}^{-1} = 1 \text{ Wb/m}^2 = 1 \text{ T}$$

An important point with respect to the above equation is that the direction of B is at right angles to both the direction of the current and the force (Figure 65). One way of remembering these directions is *Fleming's left-hand rule*.

*If the thumb, first finger and second finger of the left hand are held so that they are mutually perpendicular to each other, and the first finger points in the direction of the flux density, the second finger points in the direction of the current, then the thumb points in the direction of the **motion** of the conductor, i.e. the direction of the force.*

It is only when there is a component of the flux density which is at right angles to the direction of the current that a force will be experienced.

Example 1
What is the flux density if a conductor with a length of 2 cm in the field and carrying a current of 3 A experiences a force of 5×10^{-4} N?

$$L = 2 \text{ cm} = 2 \times 10^{-2} \text{ m}, I = 3 \text{ A}, F = 5 \times 10^{-4} \text{ N}$$

$$F = BIL$$

Hence: $$B = \frac{F}{IL}$$

$$= \frac{5 \times 10^{-4}}{3 \times 2 \times 10^{-2}} \qquad \frac{\text{N}}{\text{A} \times \text{m}}$$

$$= 8.3 \times 10^{-3} \text{ N A}^{-1} \text{ m}^{-1}$$
$$= 8.3 \times 10^{-3} \text{ T}$$
$$= 8.3 \times 10^{-3} \text{ Wb/m}^2$$

The direction of this flux density is at right angles to both the conductor and the force.

Example 2
What is the direction of the force acting on each of the current-carrying conductors shown in Figure 66?

(a) Using Fleming's left-hand rule, the force is at right angles to both B and I and directed into the paper.

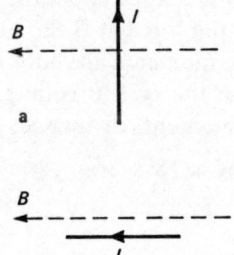

Figure 66

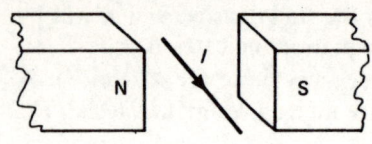

a

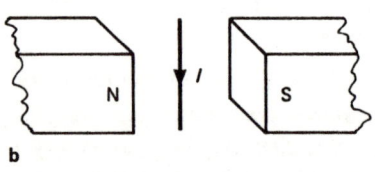

b

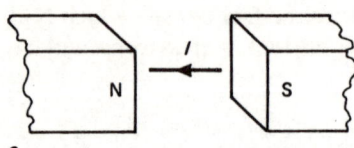

c

Figure 67

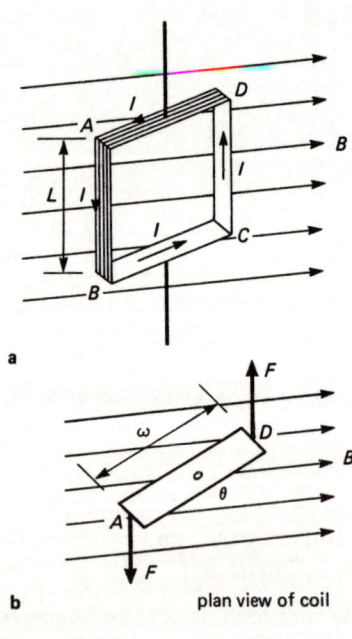

a

b   plan view of coil

Figure 68

(b)   There is no current at right angles to the magnetic flux and so there is no force acting on the conductor as a result of it being in the magnetic field.

Self-assessment questions

4   What is the flux density if a conductor with a length of 4 cm in the field and carrying a current of 2 A experiences a force of 4×10^{-3} N?

5   What is the direction of the force acting on each of the current-carrying conductors shown in Figure 67?

6   Calculate the force on a conductor of length 20 cm which carries a current of 5 A and is at right angles to a magnetic field of flux density 0.04 T.

Coils in magnetic fields

Figure 68 shows a current-carrying coil which is free to rotate about a vertical axis in a magnetic field. If you apply Fleming's left-hand rule to the currents in each of the sides, i.e. AB, BC, CD and DA, only the currents in the sides AB and CD are in such a direction to give forces which cause the coil to rotate. For these sides the flux is at right angles to them and so the force F acting on one current-carrying wire down a side is:

$$F = BIL$$

where L is the length AB or CD. If there are N wires along each side, then the force on a side is:

$$F = NBIL$$

If the coil has a width w, i.e. distance $AD = BC = w$, and the coil is at an angle θ to the field, then the perpendicular distance between the two forces is $w\cos\theta$. The perpendicular distance between the force at A and the point of rotation is thus $\frac{1}{2}w\cos\theta$. Hence the turning moment, or torque, of this force about the point of rotation is $F \times \frac{1}{2}w\cos\theta$. Similarly for the force at B the turning moment or torque is $F \times \frac{1}{2}w\cos\theta$. These moments are both in the same direction. In this case, both cause the coil to rotate in an anticlockwise direction. Thus the total moment, or torque, is:

Total moment (torque) $= (F \times \frac{1}{2}w\cos\theta) + (F \times \frac{1}{2}w\cos\theta)$
$= Fw\cos\theta$
$= NBILw\cos\theta$

This production of forces which cause rotation of a coil in a magnetic field when a current passes through it is the basis of the

Figure 69   *The basic meter construction*

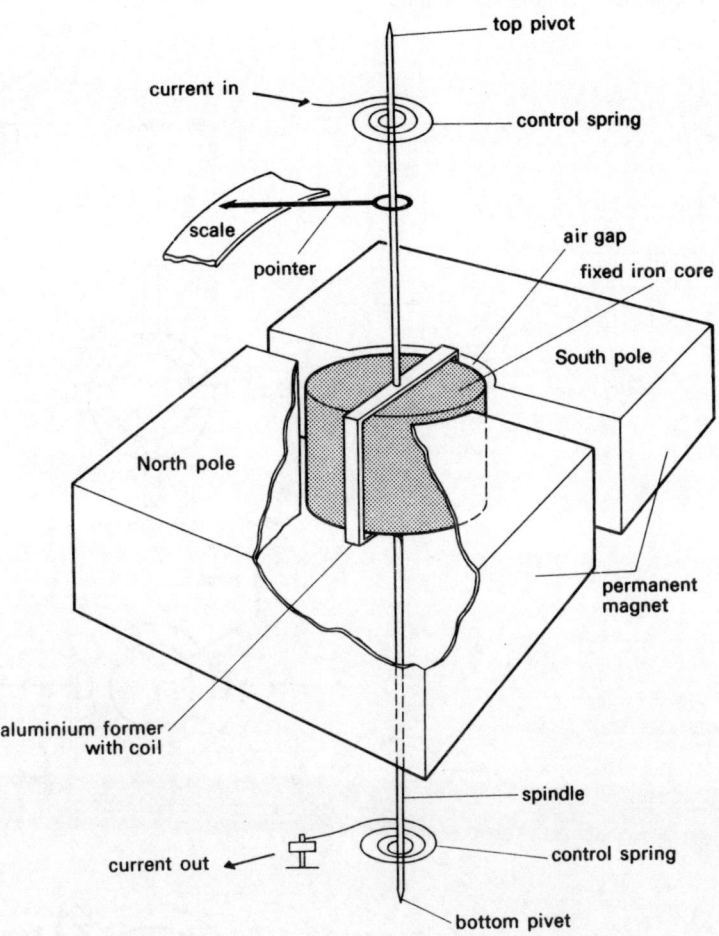

moving coil ammeters and voltmeters. Figure 69 shows the basic coil arrangement for such meters. The forces resulting from the current passing through the coil cause it to rotate against springs. The coil rotates until the moment causing the rotation is balanced by the moment exerted on the coil by the springs. The result of this rotation is a movement of a pointer across a scale to a position which depends on the current.

The production of forces which cause rotation of a coil in a magnetic field when a current passes through it is the basis of the d.c. motor. Figure 70 shows a very simple form of d.c. motor. The coil which is situated in the magnetic field is connected to an external source of d.c. through a split ring commutator and sliding contacts. When the coil rotates the commutator rotates with it but the sliding contacts remain fixed in their initial positions. The result of this is that the current through the coil is reversed every

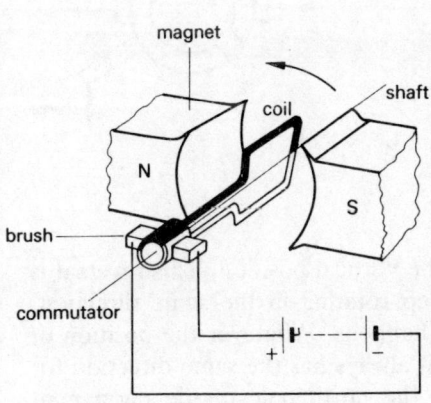

Figure 70   *A simple form of d.c. motor*

Figure 71   *End view of coil and commutator during the rotation*

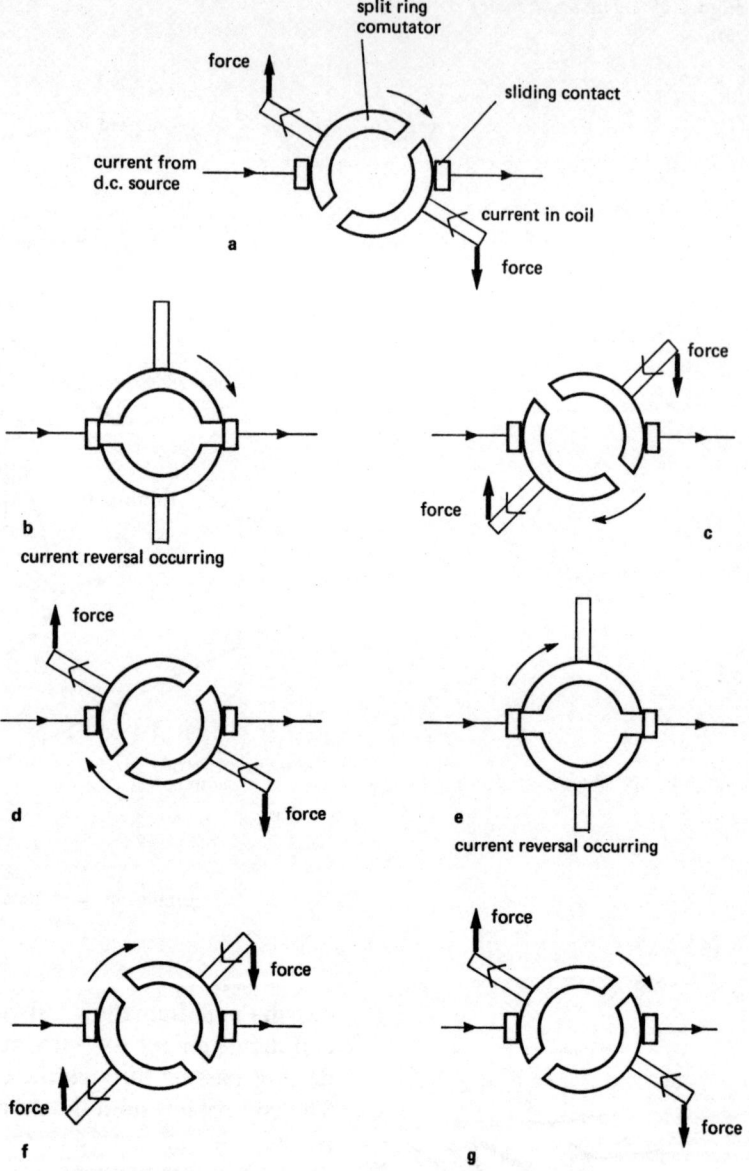

time the coil passes through the vertical position. This reversal is necessary if the coil is to keep rotating in the same direction. Figure 71 illustrates how this happens. Whatever the position of the coil, the current in the coil always has the same direction for that part of the coil near to the north pole of the permanent magnet, and the current always has the opposite direction for that part of the coil near to the south pole.

Modern motors are considerably more complex than the above, but the same principle of operation holds.

Example 3

A coil with 100 turns is wound on a rectangular frame of width 20 mm and length 25 mm. This frame is pivoted about an axis through the middle of the two shorter sides and is free to rotate about that axis. The coil is placed in a uniform magnetic field of flux density 0.5 T, the flux density being perpendicular to the axis of the coil. What is the torque on the coil when the shorter sides are inclined at an angle of (a) 0°, (b) 30°, (c) 60°, (d) 90°, to the direction of the flux density and the coil carries a current of 50 mA?

The arrangement of the coil in the magnetic field is the same as that illustrated in Figure 68.

$N = 100$, $w = 25$ mm $= 20 \times 10^{-3}$ m, $L = 25$ mm $= 25 \times 10^{-3}$ m, $B = 0.5$ T, $I = 50$ mA $= 50 \times 10^{-3}$ A

torque $= NBILw\cos \theta$

(a) $\cos \theta = \cos 0° = 1$
Hence: torque $= 100 \times 0.5 \times 0.050 \times 25 \times 10^{-3} \times 20 \times 10^{-3} \times 1$
$= 1.25 \times 10^{-3}$ N m

(b) $\cos \theta = \cos 30° = 0.866$
Hence: torque $= 0.866 \times$ the above value
$= 1.08 \times 10^{-3}$ N m

(c) $\cos \theta = \cos 60° = 0.5$
Hence: torque $= 0.5 \times$ value in (a)
$= 0.63 \times 10^{-3}$ N m

(d) $\cos \theta = \cos 90° = 0$
Hence: torque $= 0$

Self-assessment questions

7 Explain why the meter coil shown in Figure 69 rotates when a current passes through the coil. Why does it not keep on rotating with the current passing through the coil, but gives a particular amount of rotation for the current concerned?

8 Explain why the d.c. motor coil shown in Figure 70 rotates when a current passes through the coil. Why does it keep on rotating with a current passing through the coil?

9 A coil having 200 turns is wound on a rectangular frame of width 20 mm and length 30 mm. The frame is pivoted, as shown in Figure 72, so that it can rotate. The coil is in a

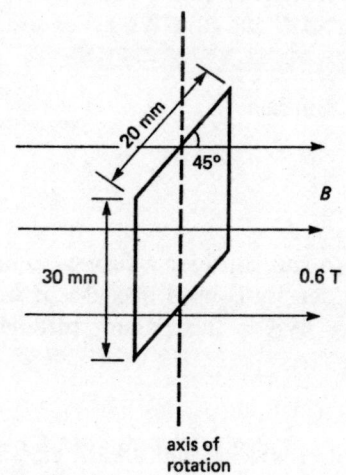

20 mm
45°
B
30 mm
0.6 T
axis of rotation

Figure 72

magnetic flux density of 0.6 T and the flux density is perpendicular to the axis of the coil. What is the torque on the coil when the shorter sides are inclined at an angle of 45° to the direction of the flux density and carry a current of 100 mA?

Magnetic flux

Figure 73 shows the type of flow pattern that occurs for a fluid, perhaps water, flowing in an orderly manner through the tube. If small pieces of paper were dropped into the water they would flow through the tube in straight lines. The pattern of flow is similar to the magnetic field pattern for a current-carrying solenoid. Within the solenoid the field lines are straight lines.

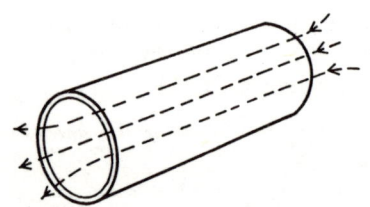

Figure 73 *Fluid flowing through a pipe*

If v is the fluid velocity and A the cross-sectional area of the tube through which the fluid flows with this velocity, then the distance covered by a particle of fluid in 1 s is v (remember that velocity is the rate at which distance is covered). The volume of fluid moved in 1 s is thus vA (see Figure 74).

rate of flow = volume of fluid moved per second
$$= vA$$

If we use the term *flux* for rate of flow, then:

flux = fluid velocity × area of cross-section

This is a general equation which refers to anything that is flowing.

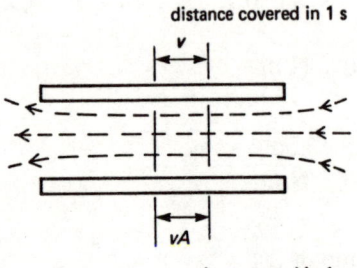

Figure 74

Because the magnetic field patterns are rather like the type of patterns that occur with fluid flow we can consider magnetism to be rather like a fluid. We thus identify a term called magnetic flux, symbol Φ and unit the weber (Wb). Instead of fluid velocity we use a term called the flux density, symbol B and unit Wb/m^2 or tesla (T). Thus:

magnetic flux Φ = flux density B × area A

$$\boxed{\Phi = BA}$$

Thus in the case of the solenoid we can consider a flux Φ to be flowing through it. The term flux density is used because, if we rearrange the above equation, $B = \Phi/A$ = flux passing through unit area.

Example 4
If the flux density in a solenoid of cross-sectional area 4 cm^2 is 0.2 T, what is the flux in the solenoid?

$B = 0.2$ T, $A = 4$ cm^2 = 4×10^{-4} m^2

$\Phi = BA$

$\quad = 0.2 \times 4 \times 10^{-4}$ $\qquad\qquad\qquad\qquad$ T $\times$ m^2

$\quad = 0.8 \times 10^{-4}$ Wb

Example 5

Calculate the magnetic flux density in a coil of cross-sectional area 5 cm^2 when a flux of 100 μWb passes through it.

$A = 5$ cm^2 = 5×10^{-4} m^2, $\Phi = 100$ μWb = 100×10^{-6} Wb

$$\Phi = BA$$

Hence: $\quad B = \dfrac{\Phi}{A}$

$$= \frac{100 \times 10^{-6}}{5 \times 10^{-4}} \qquad\qquad\qquad\qquad \frac{\text{Wb}}{\text{m}^2}$$

$$= 0.2 \text{ T}$$

Self-assessment questions

10 If the flux density in a solenoid of cross-sectional area 6 cm^2 is 0.4 T, what is the flux in the solenoid?

11 Calculate the magnetic flux density in a coil of cross-sectional area 10 cm^2 when a flux of 0.5 mWb passes through it.

Electromagnetic induction

When a permanent magnet moves in the vicinity of an electrical conductor, as shown in Figure 75, an e.m.f. is induced in it. If the magnet is stationary near the conductor then there is no e.m.f. The e.m.f. is only produced if the magnet or the conductor moves, i.e., when there is relative motion between the magnet and the conductor. The magnetic field does not have to be produced by a permanent magnet: it could be produced by a current passing through another conductor. This effect is known as electromagnetic induction.

Consider this in terms of magnetic flux. When there is relative motion between a magnet and the conductor, the conductor cuts through the magnetic flux. We can think of the magnet in Figure 75 as being like a hose pipe spraying out water: the water represents the magnetic flux. As the hose pipe moves past the conductor, the conductor has a relative motion through the water.

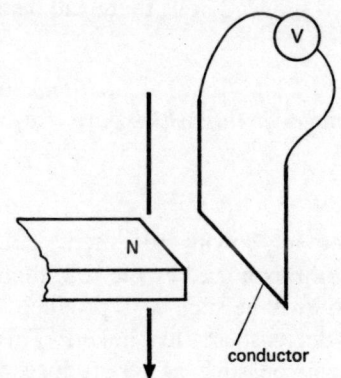

Figure 75 *Relative motion between a magnet and a conductor*

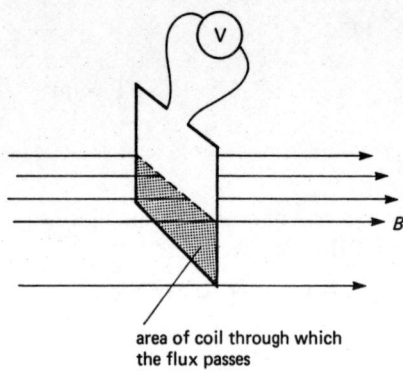

area of coil through which
the flux passes

Figure 76

An e.m.f. is induced in the conductor whenever it cuts through magnetic flux.

Figure 76 illustrates one way of considering this situation when we are concerned with a circuit, rather than a single conductor. Flux passes through the coil, in this case a single loop of wire. The shaded area in Figure 76 indicates that part of the coil area through which the flux passes. When there is relative motion between the conductor and the magnetic field, then the area of the coil through which the flux passes changes. The flux linked by the coil changes. An e.m.f. is induced when this occurs.

An e.m.f. is induced in a circuit whenever the magnetic flux linked by that circuit changes.

This is known as Faraday's first law.

The magnitude of an induced e.m.f. is directly proportional to the rate of change of flux linked by the circuit.

This is known as Faraday's second law.

Faraday's second law can be expressed as an equation which is sometimes known as Neumann's equation:

$$E = \frac{d\Phi}{dt}$$

$d\Phi/dt$ is the notation used for the rate of change with time of flux Φ linked with the coil.

If a magnet, like the one in Figure 75, moves past the loop of wire, quickly, we have a greater rate of change of flux linked than if the magnet moves slowly. The induced e.m.f. is thus greater when the magnet is moved fast than when it is moved more slowly.

When we have a circuit and an e.m.f. induced in it, there will be a current flow.

The direction of an induced current (if one flows) is such that its effect would oppose the change in magnetic flux which gave rise to it.

This is known as Lenz's law.

If the field shown in Figure 76 is moving downwards, or the coil upwards, so decreasing the amount of flux linked by the coil, then the direction of the current in the coil will be such as to produce a magnetic field which opposes the decrease in flux linked. This means that the current direction will be such as to produce a magnetic flux in the same direction as the existing flux.

If, however, the amount of flux linked by the coil were increasing, then the direction of the current in the coil would be such as to produce a magnetic field which opposed this increase. The current direction will be such as to produce a magnetic flux in the opposite direction to the existing flux.

A minus sign is often put in Neumann's equation to indicate that the induced e.m.f. would produce a current which would produce a flux change in the opposite direction to the rate of change of flux, i.e. $d\Phi/dt$, that is occurring. If $d\Phi/dt$ is increasing, the direction of the e.m.f. will produce flux to oppose the increase. If $d\Phi/dt$ is decreasing, the direction of the e.m.f. will produce a flux to oppose the decrease.

One way of arriving at the direction of the induced e.m.f. is to use Fleming's right-hand rule:

If the thumb, first and second fingers of the right hand are held mutually at right angles and the thumb points in the direction of the motion of the conductor relative to the field and the first finger points in the direction of the flux, then the second finger points in the direction of the induced current if a complete circuit is present.

Example 6
The flux linked by a single turn of wire changes from 3×10^{-3} Wb to zero in 0.2 s when a magnet moves past it. What is the average e.m.f. induced in the loop?

$$\text{flux change} = 3 \times 10^{-3} - 0 \qquad \text{Wb}$$
$$= 3 \times 10^{-3} \text{ Wb}$$

As the time taken for the change is 0.2 s then the average rate of change is:

$$\text{rate of change} = \frac{3 \times 10^{-3}}{0.2} \qquad \frac{\text{Wb}}{\text{s}}$$

$$= 1.5 \times 10^{-2} \qquad \text{Wb/s}$$
$$= 1.5 \times 10^{-2} \text{ V}$$

Note that the result is only the average induced e.m.f. because we can only calculate the average rate of change of flux. If the magnet was not moving past the coil at a steady rate, then the rate of change could vary. It is like considering a car moving along a road. We can calculate the average rate at which distance is covered, i.e. average speed, from the knowledge of the total distance covered and the total time taken. However, the speed could be quite variable during the time considered.

Example 7
The flux in a large coil is 12 mWb. If the coil has 1000 turns, what is the average induced e.m.f. when the flux decays to zero in 0.2 s?

flux change per turn = 12 mWb
$$= 12 \times 10^{-3} \text{ Wb}$$

The total flux change for the coil is the sum of the flux changes for each turn of the coil.

$$\text{total flux change} = 1000 \times 12 \times 10^{-3} \qquad \text{Wb}$$
$$= 12 \text{ Wb}$$

As the time taken for the change is 0.2 s then the average rate of change is:

$$\text{rate of change} = \frac{12}{0.2} \qquad \frac{\text{Wb}}{\text{s}}$$

$$= 60 \text{ V}$$

Example 8
What is the direction of the induced current in the loop shown in Figure 75?

The motion of the conductor relative to the field is upwards. The solution is illustrated in Figure 77: the current flows into the paper.

Example 9
Figure 78 shows a loop of wire in various positions relative to a magnet. State whether an e.m.f. would be induced in each case, and if so the direction of the resulting current.

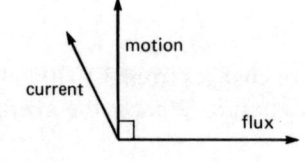

Figure 77

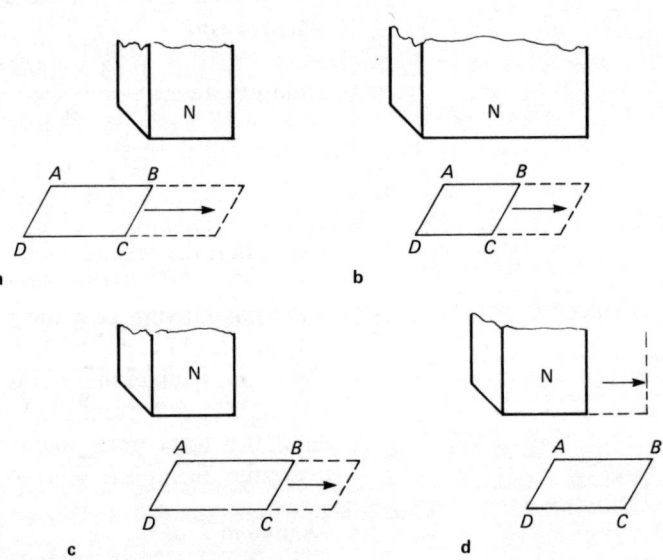

Figure 78   ABCD *is the loop of wire*

(a) The flux linked by the coil is increasing, so there will be an induced e.m.f. The current direction for the conductor in the field is *C* to *B*.

(b) There is no change in the flux linked by the coil because though there is flux linked by the coil the amount of flux is not changing. This assumes that the flux within the region immediately adjacent to the pole is constant. With such a simple situation this is unlikely to be strictly true. However, to a reasonable approximation, we must conclude that there is no induced e.m.f.

(c) The flux linked by the coil is decreasing, so there will be an induced e.m.f. The current direction for the conductor moving in the field, i.e. that part of the loop *AD*, is from *D* to *A*. This gives a current round the loop in the opposite direction to that in (a).

(d) The flux linked by the coil is increasing, so there will be an induced e.m.f. The relative motion of the conductor *AD* is to the left, so the current direction in it is from *A* to *D*.

Note: In the above questions Fleming's right-hand rule was used to determine the current direction. To use this rule with the loop of wire you really have to consider each of the conductors *AB*, *BC*, *CD* and *DA*. However, the conductors *AB* and *CD* are not moving in such a direction that a current at right angles to both the motion and the field can be produced. In (a) only the conductor *BC* is moving in the field; in (c) and (d) the only conductor moving in such a direction that a current can flow along it is *AD*. In the case of (b) we can argue that there will be no induced e.m.f. because there is no change in the flux linked. An alternative is to apply Fleming's right-hand rule to both *AD* and *BC*. The direction of the current in each is in the opposite direction, so if each is cutting flux at the same rate, then the currents will cancel each other out and there will be zero current.

Self-assessment questions

12 Explain what is meant by electromagnetic induction.

13 State Faraday's laws and Lenz's law.

14 Describe how you could demonstrate electromagnetic induction in the laboratory.

15 The flux linked by a single turn of wire changes from 0.5 mWb to zero in 0.1 s. What is the average e.m.f. induced in the loop?

16 The flux linked by a coil of 400 turns changes from 6 mWb to zero in 0.3 s. What is the average induced e.m.f.?

17   The flux linked by a coil of 300 turns changes from zero to 2 mWb in 0.1 s. What is the average induced e.m.f.?

18   Figure 79 shows a loop of wire in various positions relative to a magnet. State whether an e.m.f. would be induced in each case, and if so the direction of the resulting current.

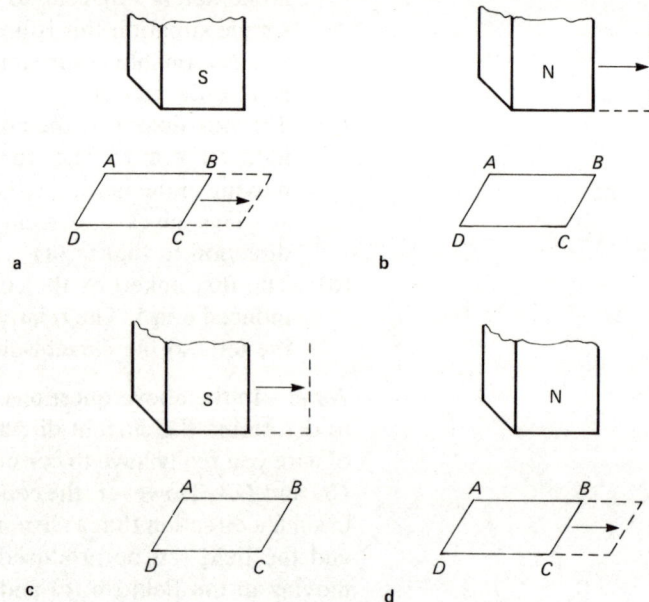

Figure 79   ABCD *is the loop of wire*

The induced e.m.f.

Figure 80 shows a single turn coil of wire moving through a magnetic field and the flux linked by the coil is changing. The coil is moving with a constant velocity in a direction at right angles to the flux density. The result is an induced e.m.f. and a current in the direction shown. Figure 80(a) shows the position of the coil at one time and Figure 80(b) shows its position a small time interval of Δt later (note that when Δ (delta) appears in front of a symbol it means a small increment of it). Initially the flux linked by the loop is that flux which passes through the marked area in (a). After a time Δt this area has increased. It has increased because the loop is moving with a velocity v through the magnetic field. Moving with this velocity, it covers a distance of $v\Delta t$ in the time. Thus the increase in area is $Lv\Delta t$. Since the flux density is B in a direction at right angles to the coil, the change in flux in the time Δt is $B \times$

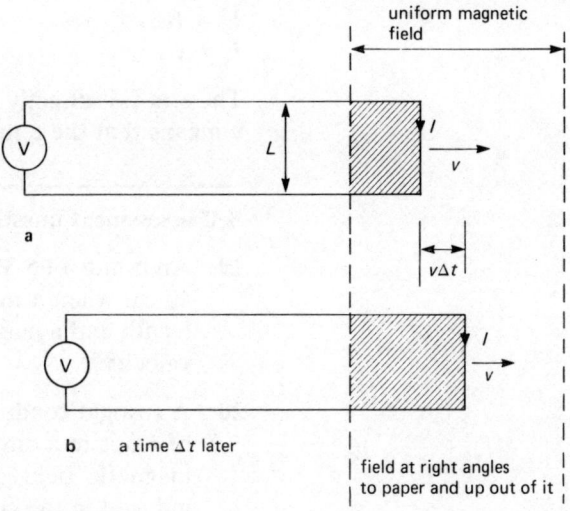

Figure 80

change in area = $BLv\Delta t$. But the rate of change of flux with time is the induced e.m.f. hence:

$$E = \frac{BLv\Delta t}{\Delta t}$$

$$\boxed{E = BLv}$$

E is the value of the e.m.f. at a time when the coil is moving with a velocity v.

Example 10
A straight conductor of length 200 mm moves with a velocity of 6 m/s in a direction perpendicular to both the magnetic field and the conductor. If the field has a flux density of 0.3 T, what is the e.m.f. induced in the conductor?

$L = 200$ mm $= 200 \times 10^{-3}$ m, $v = 6$ m/s, $B = 0.3$ T
$E = BLv$
   $= 0.3 \times 200 \times 10^{-3} \times 6$ T × m/s × m
   $= 0.36$ V

Example 11
A straight conductor moves with a constant velocity in a direction perpendicular to both the magnetic field and the conductor. If the velocity is doubled, how does the induced e.m.f. change?

$$E = BLv$$
$$E \propto v$$

The e.m.f. is directly proportional to the velocity v. Thus doubling v means that the e.m.f. is doubled.

Self-assessment questions

19   An e.m.f. of 5 V is induced in a straight conductor of length 40 cm when it moves in a direction perpendicular to both its length and a magnetic field of flux density 0.3 T. What is the velocity?

20   A straight conductor of length 0.4 m is moved with a velocity of 4 m/s in a direction perpendicular to both its length and a magnetic field of flux density 0.4 T. What is the e.m.f. induced in the conductor?

21   A straight conductor moves with a constant velocity in a direction perpendicular to both the magnetic field and the conductor.
(a)   If the length of the conductor is doubled how does the e.m.f. induced in the conductor change?
(b)   If the flux density is doubled, how does the e.m.f. induced in the conductor change?

The simple a.c. generator

The amount of flux linked by a coil depends on its angle relative to a magnetic field. If the coil is at right angles to the field, as shown in Figure 81(a), then a maximum amount of flux is linked. If, however, the coil is in the same plane as the field, then no flux is linked (see Figure 81(b)). Thus if a coil is rotated in a magnetic

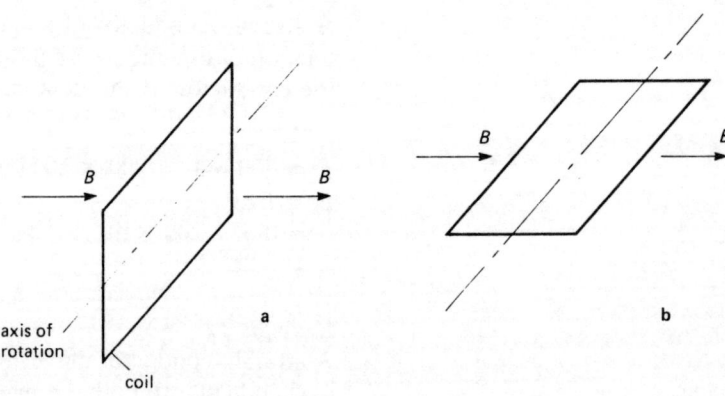

Figure 81   (a) *Maximum flux linked*
(b) *Zero flux linked*

Figure 82 *A simple a.c. generator*

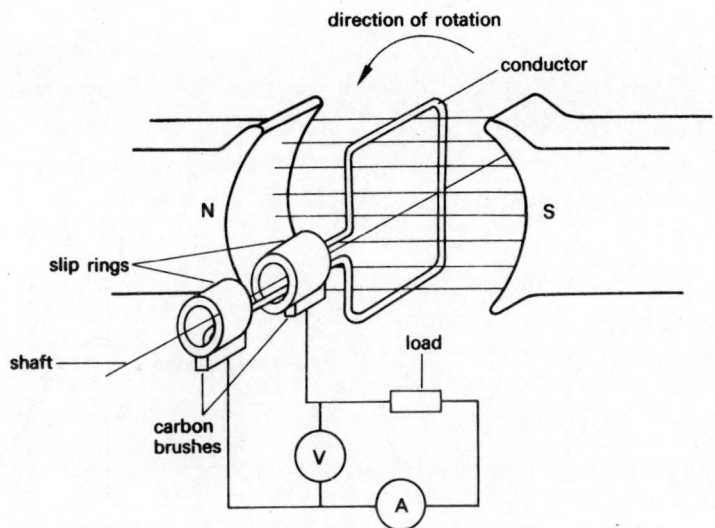

field of constant direction, then the flux linked by that coil will vary. A changing flux linkage means an induced e.m.f. This is the basis of the simple a.c. generator.

Figure 82 shows a simple a.c. generator. It consists of a coil in a magnetic field. The coil is shown as a turn of wire wound on an armature. This armature is rotated at a constant angular speed. Thus the flux linked by the coil will be continually changing. An e.m.f. is induced. The coil is connected to sliding contacts via slip rings, which rotate with the drive shaft, and hence give an output. The output is an alternating current.

Figure 83 shows how the alternating current results from the changes in flux linkage by the rotating coil. For coil position 1 in relation to the field direction, the maximum flux is linked. However, when the position of the coil changes slightly, the amount of flux linked barely changes. Thus the rate of change of flux linked by the rotating coil when it is moving through position 1 is zero. The induced e.m.f. is thus zero. When the coil has moved to position 2 the flux linked has changed to zero. However, a small change in the position of the rotating coil from this position considerably affects the amount of flux linked. At this position the rate of change of flux linked by the rotating coil is a maximum, the induced e.m.f. is thus also a maximum. When the coil continues rotating to position 3, the flux linked again increases to a maximum and the rate of change of flux linked decreases to zero. When the coil rotates to position 4 the flux linked again drops to zero and the rate of change of flux linked is a maximum. The e.m.f. induced thus changes from zero at position 1, to a maximum

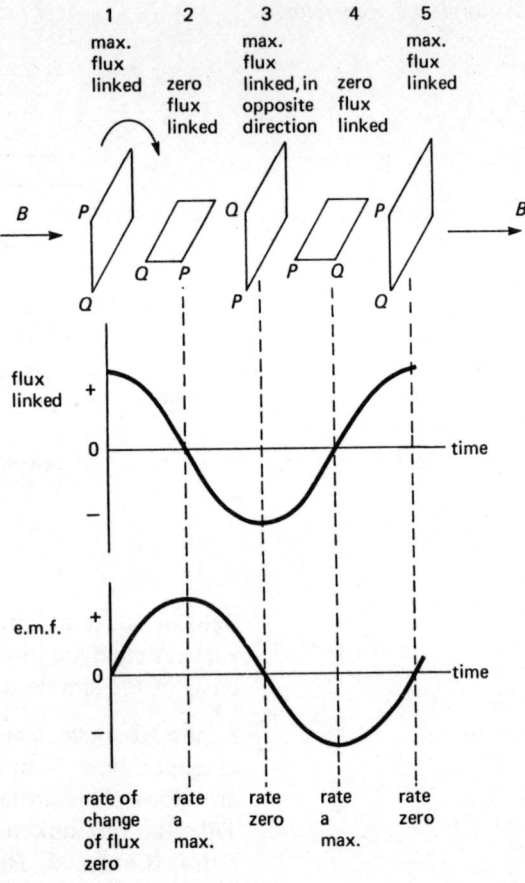

Figure 83

at position 2, to zero at position 3, to a maximum at position 4, to zero at position 5. Because the coil is rotating, the flux passing through it when it is position 3 is in the opposite direction to that in position 1. The effect of this is to give an alternating e.m.f. which oscillates from positive to negative values.

Self-assessment questions

22   Explain how a rotating coil in a magnetic field is able to generate an alternating e.m.f.

23   At what angle with respect to the direction of the magnetic field will the induced e.m.f. be (a) a maximum (b) zero, for the rotating coil illustrated in Figure 82?

4 Magnetic circuits and materials

After reading this chapter you should be able to:

1. Compare magnetic circuits with electrical circuits, explaining the terms m.m.f., reluctance, flux and permittivity.
2. Use the equations m.m.f. $= \Phi S$ and $S = L/(\mu A)$.
3. Solve problems involving series connected magnetic circuits, involving not more than one change in dimension or material.
4. Explain the effects of flux leakage and fringing.
5. Explain the term magnetic field intensity and use the equation $B = \mu H$.
6. Distinguish between ferromagnetic, paramagnetic and diamagnetic materials.
7. Describe magnetization curves for ferromagnetic materials and how the relative permeability varies with magnetic field intensity.
8. Solve magnetic circuit problems using B–H graphs.
9. Explain the significance of hysteresis loops.
10. Explain magnetic screening.

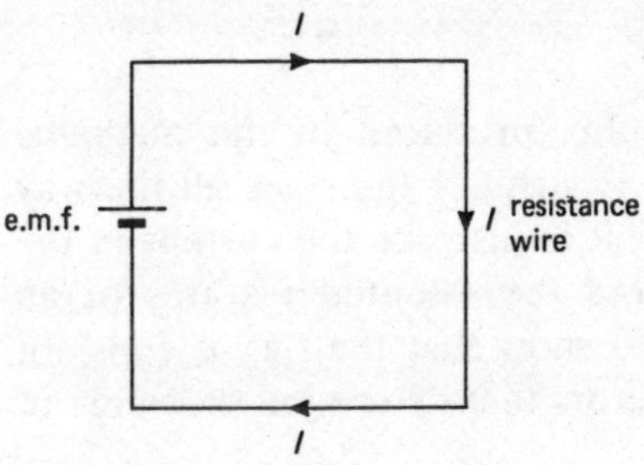

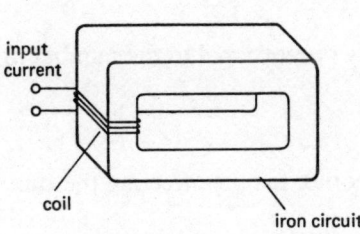

Figure 84   *A simple electrical circuit*

Magnetic circuits

A simple electrical circuit can be considered to consist of a source of e.m.f., perhaps a cell, with a resistive load connected across the terminals of the cell. This resistive load might be just a length of resistance wire. Figure 84 shows this type of circuit. At every position round this circuit, a series circuit, the current is the same.

Figure 85 shows a simple magnetic circuit. It consists of a closed circuit made of iron, which is a magnetic material. In the electrical circuit an electrical current flowed, whereas in the magnetic circuit there is a magnetic flux. The source of this magnetic flux is a coil of wire wrapped round the iron and supplied with a current. The coil of wire with the current in it behaves like the source of the e.m.f. in the electrical circuit. Indeed it is convenient to refer to the magnetic source as a source of m.m.f. (e.m.f. stands for electromotive-force, m.m.f. stands for **magnetomotive-force**).

Figure 85   *A simple magnetic circuit*

Figure 86 *Investigating the factors determining the flux in a magnetic circuit. The input is an alternating current in a coil. This gives rise to an alternating flux which then induces an e.m.f. in the flux measuring coil. The induced e.m.f. results in a trace on an oscilloscope screen, the height of the trace being proportional to the maximum flux linked*

(a) When the flux measuring coil is moved to different positions round the magnetic circuit, no change in the output seen on the oscilloscope screen occurs. The flux linked by the coil is thus constant round the circuit

(b) Changing the current in the coil and changing the number of turns in the coil affect the flux measured in the circuit. $\Phi \propto NI$

(c) Changing the length of the magnetic flux path and changing the cross-sectional area of the flux path affect the flux measured in the circuit. $\Phi \propto A/L$

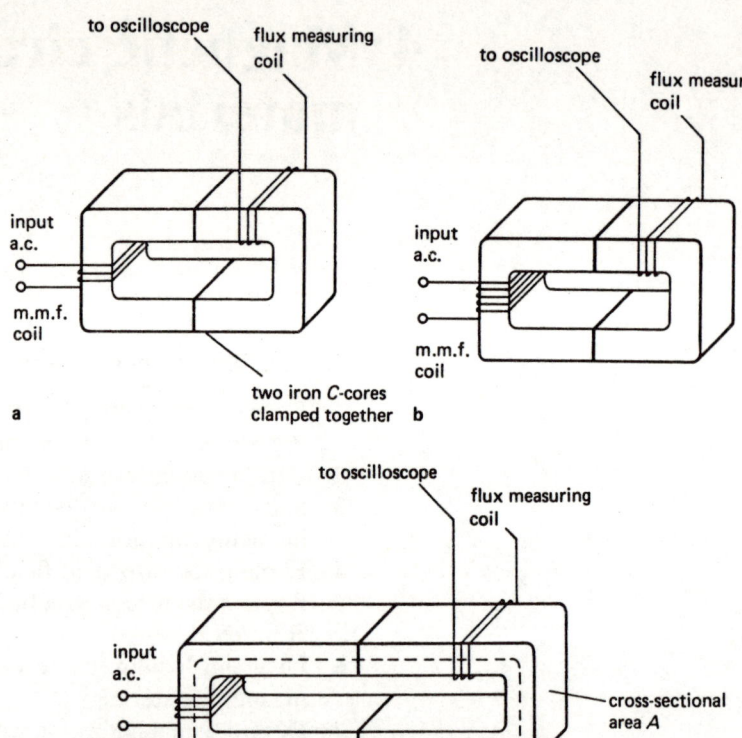

With the magnetic circuit the flux produced in the magnetic material by a current in the coil is virtually the same all the way round the circuit. In this respect it is just like the current in the electrical circuit. Figure 86 shows the essential features of an experiment that could be used to show that the flux is constant round the circuit and also the factors that determine the value of the flux.

For a given circuit, the flux is found to be directly proportional to the current I in the input coil.

$$\Phi \propto I$$

The flux is also found to be directly proportional to the number of turns N on the coil.

$$\Phi \propto N$$

For the given circuit there are no other factors affecting the flux, so we have:

$$\Phi \propto NI$$

Now in the case of an electrical circuit, a change in N and I is equivalent to keeping the circuit constant and changing the cell, i.e. the e.m.f. Hence we define the m.m.f. as being NI.

$$\boxed{\text{m.m.f.} = NI}$$

The usual unit of the m.m.f. is the ampere. Sometimes the unit ampere turns is used.

The current in the simple electrical circuit can be changed by changing the circuit resistance, keeping the e.m.f. constant. This resistance change can be achieved by changing the length of the current path and the cross-sectional area of the conductor. With the magnetic circuit we can do the same thing. The flux in the circuit is found to depend on the length of the flux path and the cross-sectional area of the flux path. For this reason we can use a term similar in concept to that of electrical resistance. This term is reluctance, symbol S. With electrical resistance we can define it as being $R = V/I$, with reluctance we can define it as being m.m.f./Φ.

$$S = \frac{\text{m.m.f.}}{\Phi}$$

or $$\boxed{\text{m.m.f.} = \Phi S}$$

Compare this with $V = IR$. The unit of reluctance is A/Wb.

For the magnetic circuit used in Figure 86, the flux in the circuit for a given m.m.f. is found to be proportional to the cross-sectional area A of the flux path and inversely proportional to the length L of the flux path.

$$\Phi \propto A$$
$$\Phi \propto \frac{1}{L}$$

Hence: $$\Phi \propto \frac{A}{L}$$

In an electrical circuit the current in a resistance element depends not only on the cross-sectional area of that element and its length but also on the type of material used. This is also true in the case of a magnetic circuit. The flux depends on the material used for the magnetic circuit. For an electrical circuit we refer to the conductivity (or resistivity) of a material. The equivalent for a magnetic circuit is the absolute permeability μ of the material.

Hence we have:

$$\Phi \propto \frac{\mu A}{L}$$

The bigger the permeability, the bigger the flux. The bigger the quantity $\mu A/L$, the bigger the flux. However the bigger the flux, the smaller the reluctance (remember $S = $ m.m.f.$/\Phi$). Hence we have:

$$S = \frac{L}{\mu A}$$

Note: This equation is comparable to the electrical equation of $R = \rho L/A$, where ρ is the resistivity, or the equation $R = L/(\sigma A)$, where σ is the conductivity.

The permeability μ is often written in another form:

$$\mu = \mu_0 \mu_r$$

μ_0 is known as the permeability of free space and μ_r the relative permeability. In the case of a vacuum (or air which can be taken as near approximation) the relative permeability is 1. Thus μ_0 is the value of the permeability for a vacuum and μ_r is the factor by which this permeability must be multiplied when a material is involved. μ_0 has the value of $4\pi \times 10^{-7}$ T m A^{-1}. An alternative equivalent unit is H/m, where H stands for henry. This unit will be met in the next chapter.

The relative permeability of iron can be many hundreds. In other words having an iron circuit rather than just air increases the flux, for the same m.m.f., by a factor of hundreds. A simple example of this is that an air-cored solenoid is a much weaker electromagnet than an iron-cored solenoid.

Example 1
A magnetic circuit can be compared with an electrical circuit. What are the magnetic circuit quantities which correspond with (a) electrical current, (b) e.m.f., (c) resistance?

(a) Electrical current – flux.
(b) E.m.f. – m.m.f.
(c) Resistance – reluctance.

Example 2
A coil has 200 turns and carries a current of 0.5 A. What is its m.m.f.?

$N = 200$, $I = 0.5$ A

m.m.f. = NI

$\qquad$ = 200×0.5 $\hfill$ A

$\qquad$ = 100 A

Example 3

Calculate the reluctance of a magnetic circuit which has a coil supplying a m.m.f. of 200 A and has a flux in the magnetic core of 300 µWb.

m.m.f. = 200 A, $\Phi = 300$ µWb = 300×10^{-6} Wb

$\qquad$ m.m.f. = ΦS

Hence: $\quad S = \dfrac{\text{m.m.f.}}{\Phi}$

$$= \frac{200}{300 \times 10^{-6}} \qquad \frac{\text{A}}{\text{Wb}}$$

$$= 6.67 \times 10^{5} \text{ A/Wb}^{-1}$$

Example 4

A material has an absolute permeability of 8×10^{-5} T m A^{-1}. What is its relative permeability?

$\mu_0 = 4\pi \times 10^{-7}$ T m A^{-1}, $\mu = 8 \times 10^{-5}$ T m A^{-1}

$\qquad \mu = \mu_0 \, \mu_r$

Hence: $\quad \mu_r = \dfrac{\mu}{\mu_0}$

$$= \frac{8 \times 10^{-5}}{4\pi \times 10^{-7}} \qquad \frac{\text{T m A}^{-1}}{\text{T m A}^{-1}}$$

$$= 63.7$$

Example 5

Calculate the reluctance of a magnetic circuit with a mean flux path length of 0.2 m, a cross-sectional area of 0.002 m^2 and a relative permeability of 120.

$L = 0.2$ m, $A = 0.002$ m^2, $\mu_r = 120$, $\mu_0 = 4\pi \times 10^{-7}$ T m A^{-1}

$$S = \frac{L}{\mu A}$$

$$= \frac{L}{\mu_r \, \mu_0 A}$$

$$= \frac{0.2}{120 \times 4\pi \times 10^{-7} \times 0.002} \qquad \frac{m}{T\,m\,A^{-1} \times m^2}$$

$$= 6.63 \times 10^5 \text{ A/Wb}$$

Example 6

Determine the flux in the core of a magnetic circuit with a mean flux path length of 400 mm, a cross-sectional area of 300 mm² and a relative permeability of 100, when it is supplied with a m.m.f. of 240 A.

$L = 400$ mm $= 400 \times 10^{-3}$ m, $A = 300$ mm² $= 300 \times 10^{-6}$ m², $\mu_r = 100$, $\mu_0 = 4\pi \times 10^{-7}$ T m A^{-1}, m.m.f. $= 240$ A

$$\text{m.m.f.} = \Phi S$$

Hence: $\quad \Phi = \dfrac{\text{m.m.f.}}{S}$

But: $\quad S = \dfrac{L}{\mu_0 \mu_r A}$

Hence: $\quad \Phi = \dfrac{\text{m.m.f.} \times \mu_0 \mu_r A}{L}$

$$= \frac{240 \times 4\pi \times 10^{-7} \times 100 \times 300 \times 10^{-6}}{400 \times 10^{-3}} \qquad \frac{A \times T\,m\,A^{-1} \times m^2}{m}$$

$$= 2.26 \times 10^{-5} \text{ Wb}$$

Self-assessment questions

1 Define the terms (a) m.m.f. (b) reluctance and (c) relative permeability.

2 A magnetic circuit can be compared with an electrical circuit. What are the electric circuit quantities which correspond to (a) flux (b) m.m.f. (c) reluctance?

3 A coil has 400 turns and carries a current of 300 mA. What is its m.m.f.?

4 A 200 turn coil is required to generate a m.m.f. of 200 A. What should be the current in the coil?

5 Calculate the reluctance of a magnetic circuit which has a coil supplying a m.m.f. of 300 A and has a flux in the magnetic core of 400 μWb.

6 What will be the flux in the core of a magnetic circuit if it has a reluctance of 2×10^5 A/Wb and a m.m.f. of 200 A is used?

7 What is the equivalent of the electrical circuit equation $V = IR$ in the magnetic circuit?

8 What is the relative permeability of a material with an absolute permeability of 1.2×10^{-4} T m A^{-1}?

9 What is the absolute permeability of a material with a relative permeability of 90?

10 Calculate the reluctance of a magnetic circuit with a mean flux path length of 0.3 m, a cross-sectional area of 20 cm^2 and a relative permeability of 100.

11 Calculate the flux in the core of a magnetic circuit with a mean flux path length of 300 mm, a cross-sectional area of 200 mm^2 and a relative permeability of 80 when it is supplied with a m.m.f. of 300 A.

12 What is the m.m.f. required to produce a flux of 500 μWb in a magnetic circuit with a flux path length of 400 mm, a cross-sectional area of 30 cm^2 and a relative permeability of 120?

13 State the SI units of (a) m.m.f. (b) flux (c) reluctance and (d) absolute permeability.

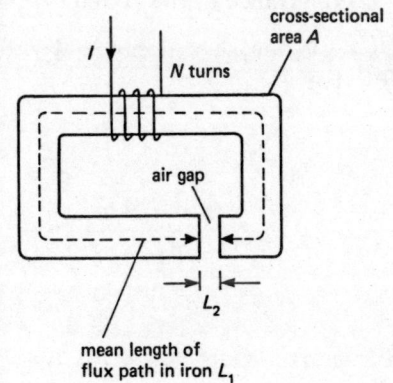

Figure 87 *A series connected magnetic circuit*

Series connected magnetic circuits

In a series connected electrical circuit the current is the same in all components and the total circuit resistance is obtained by adding together the resistances of each of the circuit components.

$$R = R_1 + R_2 + R_3$$

In a series connected magnetic circuit the flux is the same in all components.

Thus for the circuit shown in Figure 87 the components are a length of iron and an air gap. We can consider this to be a circuit with two reluctances in series, the reluctance of the iron and the reluctance of the air gap.

With the electrical circuit the way of arriving at the above equation for the total circuit resistance was to consider that the total e.m.f. was equal to the sum of the potential differences across each resistor in the circuit (see Chapter 1). With the magnetic circuit the m.m.f. supplied by the coil is partly across one reluctance and partly across the other such that:

m.m.f.$_{total}$ = m.m.f.$_{iron}$ + m.m.f.$_{air}$

But as m.m.f. = ϕS, we have:

$$\phi S_{total} = \phi S_{iron} + \phi S_{air}$$

Hence: $S_{total} = S_{iron} + S_{air}$

In general we can write the equation for the total reluctance S of reluctances S_1, S_2, S_3, etc. in series:

$$\boxed{S = S_1 + S_2 + S_3}$$

For the example shown in Figure 87 we have for each reluctance:

$$S_{iron} = \frac{L_1}{\mu_{iron}A}$$

$$S_{air} = \frac{L_2}{\mu_{air}A}$$

The areas, in this particular example, are the same.

Example 7

A magnetic circuit consists of an iron core, of mean length 180 mm, with an air gap of mean length 20 mm. The cross-sectional area for the flux path is 4 cm^2 for both the iron and the air. If the iron has a relative permeability of 120 and that of air is assumed to be 1, what is the total reluctance of the circuit?

For the iron: L = 180 mm = 180×10^{-3} m, A = 4 cm^2 = 4×10^{-4} m^2, μ_r = 120, $\mu_0 = 4\pi \times 10^{-7}$ T m A^{-1}

$$S = \frac{L}{\mu_r \mu_0 A}$$

$$= \frac{180 \times 10^{-3}}{120 \times 4\pi \times 10^{-7} \times 4 \times 10^{-4}} \qquad \frac{m}{T\,m\,A^{-1} \times m^2}$$

$$= 2.98 \times 10^6 \text{ A/Wb}$$

For the air: L = 20 mm = 20×10^{-3} m, A = 4 cm^2 = 4×10^{-4} m, μ_r = 1, $\mu_0 = 4\pi \times 10^{-7}$ T m A^{-1}

$$S = \frac{L}{\mu_r \mu_0 A}$$

$$= \frac{20 \times 10^{-3}}{1 \times 4\pi \times 10^{-7} \times 4 \times 10^{-4}} \qquad \frac{m}{T\,m\,A^{-1} \times m^2}$$

$$= 3.98 \times 10^7 \text{ A/Wb}$$

Note: Even though the air gap is smaller than the iron, it has a larger reluctance. This is because of the lower permeability.

Total reluctance $S = S_1 + S_2$
$$= 2.98 \times 10^6 + 3.98 \times 10^7 \qquad \text{A/Wb}$$
$$= 4.28 \times 10^6 \text{ A/Wb}$$

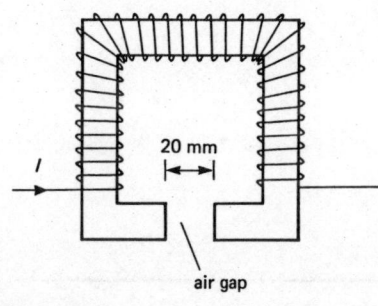

Figure 88

Example 8
An electromagnet has an iron core of length 200 mm and is wound with a coil of 500 turns (see Figure 88). When the current through the coil is 2 A, what is the flux density in the air gap between the poles of the magnet? The air gap has a length of 20 mm and the relative permeability of the iron may be taken as 1000.

For the iron: $L = 200$ mm $= 200 \times 10^{-3}$, $\mu_r = 1000$, $\mu_0 = 4\pi \times 10^{-7}$ T m A^{-1}

$$S = \frac{L}{\mu_r \mu_0 A}$$

$$= \frac{200 \times 10^{-3}}{1000 \times 4\pi \times 10^{-7} \times A}$$

$$= \frac{159}{A}$$

For the air: $L = 20$ mm $= 20 \times 10^{-3}$ m, $\mu_r = 1$, $\mu_0 = 4\pi \times 10^{-7}$ T m A^{-1}

$$S = \frac{L}{\mu_r \mu_0 A}$$

$$= \frac{20 \times 10^{-3}}{1 \times 4\pi \times 10^{-7} \times A}$$

$$= \frac{159 \times 10^2}{A}$$

If we assume that the area A is the same for both the iron and the air gap, then:

total reluctance $S = S_1 + S_2$
$$= \frac{159}{A} + \frac{159 \times 10^2}{A}$$

$$= \frac{159 + (159 \times 10^2)}{A}$$

$$= \frac{161 \times 10^2}{A}$$

For the circuit, the m.m.f. $= NI = 500 \times 2 = 1000$ A.
But: m.m.f. $= \Phi S$

Hence: $\Phi = \dfrac{\text{m.m.f.}}{S}$

$$= \frac{1000}{(161 \times 10^2/A)}$$

$$= 0.0621 \ A$$

The flux density $B = \Phi/A$, hence:

$B = \dfrac{0.0621A}{A}$

$= 0.0621$ Wb/m^2

Self-assessment questions

14   In a series electrical circuit the current in each component is the same. What is the same in a series magnetic circuit?

15   A magnetic circuit consists of an iron core with a reluctance of 3×10^6 A/Wb and an air gap with a reluctance of 40×10^6 A/Wb.
   (a)   What is the total reluctance of the circuit?
   (b)   If there is a m.m.f. of 200 A what is the flux in the air gap?

16   A magnetic circuit consists of an iron core of mean length 200 mm, with an air gap of length 15 mm. The cross-sectional area for the flux path of both the iron and the air is 6 cm^2. If the iron has a relative permeability of 400 and that of the air is assumed to be 1, what is the total reluctance of the circuit?

17   An electromagnet has an iron core of length 300 mm and is wound with a coil of 300 turns. When the current through the coil is 2 A, what is the flux density in the air gap between the poles of the electromagnet? The air gap has a length of 30 mm, the relative permeability of the iron is 800 and the cross-sectional areas of the iron and air may be considered to be the same. The electromagnet is of the form illustrated in Figure 88.

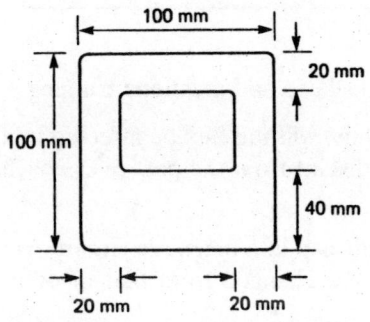

Figure 89

18 A magnetic circuit consists of two parts in series. One part is iron of cross-sectional area 4×10^{-3} m^2 and length 220 mm, the other part the same material but with a cross-sectional area of 2×10^{-3} m^2 and length 100 mm.

(a) What is the total reluctance of the circuit if the iron has a permeability of 120?

(b) What is the flux in the circuit when an m.m.f. of 200 A is applied?

19 Calculate the reluctance of the magnetic circuit shown in Figure 89. The mild steel has a permeability of 500 and the material has a constant thickness of 10 mm.

20 An iron ring has a cross-sectional area of 300 mm^2 and a mean diameter of 350 mm is wound with 800 turns of wire. If the relative permeability of the iron is taken as 800, what is the flux in the iron when a current of 300 mA flows through the coil? A radial air gap 0.5 mm wide is then cut in the ring. What is the flux in the iron now?

Flux leakage and fringing

In all the questions considered so far for magnetic circuits it has been assumed that the flux everywhere in the series connected circuit is the same, even when there are breaks in the iron and air gaps are present. In other words, flux behaves like water flowing through a pipe and it is assumed that none of it leaks from the pipe. This is difficult to achieve in practice. Invariably the flux produced inside the coil by the m.m.f. is not the same as the flux in some other part of the magnetic circuit.

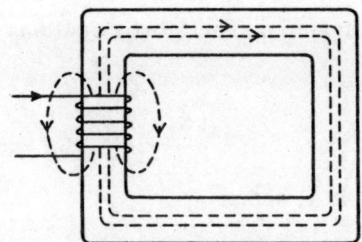

Figure 90 *Magnetic leakage*

Figure 90 shows the effect known as magnetic leakage. Not all the flux which links the coil passes right round the circuit. The result is that the flux within the coil is greater than the flux in other parts of the circuit.

Figure 91 shows the effect known as magnetic fringing. The flux passing through an air gap tends to spread out. This does not affect the flux passing through that part of the circuit but does mean that the flux density is reduced within the air gap because of this spreading out of the flux over a greater area.

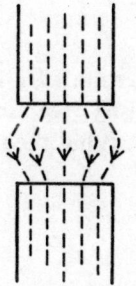

Figure 91 *Magnetic fringing*

In both magnetic leakage and magnetic fringing, correction factors can be used to allow for such effects and so enable the flux and flux density to be determined.

Self-assessment questions

21 Explain the terms magnetic leakage and magnetic fringing.

22 If magnetic leakage occurs, how will the flux be affected in a part of the magnetic circuit distant from the part over which the coil is wound?

23 When fringing occurs in an air gap in a magnetic circuit, will the (a) flux (b) flux density, be changed from that in other parts of the circuit?

Magnetic field intensity

For a magnetic circuit we have the general relationship

m.m.f. $= \Phi S$

The reluctance can be expressed in terms of the length of the magnetic circuit L and the cross-sectional area of that circuit material.

$$S = \frac{L}{\mu A}$$

where μ is the permeability of the material. These two equations can be combined to give:

$$\text{m.m.f.} = \Phi \times \frac{L}{\mu A}$$

The flux density B in the circuit is Φ/A, hence:

$$\text{m.m.f.} = B \times \frac{L}{\mu}$$

Rearranging this equation gives:

$$B = \mu \times \frac{\text{m.m.f.}}{L}$$

The m.m.f. per unit length of the circuit, i.e. m.m.f./L, is given a special name – the magnetic field intensity (or sometimes magnetizing force or magnetic field strength). It has the symbol H and the unit A/m.

$$H = \frac{\text{m.m.f.}}{L}$$

Hence the equation becomes:

$$B = \mu H = \mu_r \mu_0 H$$

Permeability can be defined as the flux density divided by the magnetic field intensity.

We can compare the magnetic field intensity, i.e. m.m.f./L, to the potential difference per unit length of a current-carrying conductor in an electrical circuit. Before the current in such an electrical conductor can be calculated from a knowledge of the potential difference per unit length, the electrical properties of the material used as the conductor have to be taken into account. Thus a length of copper conductor would give rise to a different current from a length of aluminium because they have different resistivities. A similar situation occurs with a magnetic circuit.

A coil producing a m.m.f. gives rise to a m.m.f. per unit length of a magnetic circuit, i.e. a magnetic field intensity. This will occur regardless of the material used for the magnetic circuit. The flux and the flux density that are produced within the material will, however, depend on the magnetic properties of the material. This is represented by the permeability.

Example 9
A coil of length 150 mm has 800 turns and carries a current of 2 A. What is the magnetic field intensity produced within the coil?

$L = 150$ mm $= 150 \times 10^{-3}$ m, $N = 800$, $I = 2$ A

$$H = \frac{\text{m.m.f.}}{L}$$

But: m.m.f. $= NI$

Hence: $$H = \frac{NI}{L}$$

$$= \frac{800 \times 2}{150 \times 10^{-3}} \qquad \frac{\text{A}}{\text{m}}$$

$$= 1.07 \times 10^4 \text{ A/m}$$

Example 10
A coil having 1200 turns is wound on a ring of mean diameter 120 mm. What is the m.m.f. produced by the coil and the magnetic field intensity in the coil when a current of 2 A passes through it?

$N = 1200$, $d = 120$ mm $= 120 \times 10^{-3}$ m, $I = 2$ A

m.m.f. $= NI$

$\qquad = 1200 \times 2$ \hfill A

$\qquad = 2400$ A

$$L = \pi d = \pi \times 120 \times 10^{-3} \text{ m}$$

Hence: $H = \dfrac{\text{m.m.f.}}{L}$

$$= \dfrac{2400}{\pi \times 120 \times 10^{-3}} \qquad\qquad \dfrac{\text{A}}{\text{m}}$$

$$= 6.37 \times 10^3 \text{ A/m}$$

Example 11

A coil produces a magnetic field intensity of 5000 A/m. What will be the flux density produced in an iron core having a relative permeability of 500?

$H = 5000$ A/m, $\mu_r = 500$, $\mu_0 = 4\pi \times 10^{-7}$ T m A^{-1}

$$\begin{aligned}
B &= \mu_r \mu_0 H \\
&= 500 \times 4\pi \times 10^{-7} \times 5000 \qquad\qquad \text{T m A}^{-1} \times \text{(A/m)} \\
&= 3.14 \text{ T}
\end{aligned}$$

Self-assessment questions

24 What is the magnetic field intensity in a coil of length 200 mm and 400 turns when it carries a current of 400 mA?

25 A coil of 900 turns is wound on a ring of mean diameter 100 mm. What is the m.m.f. produced by the coil and the magnetic field intensity in the coil when a current of 500 mA passes through it?

26 A coil produces a magnetic field intensity of 1000 A/m. What will be the flux density produced in an iron core with a relative permeability of 150?

Magnetic materials

Depending on the value of the relative permeability, a material can be considered to be ferromagnetic, paramagnetic or diamagnetic. Ferromagnetic materials have relative permeabilities considerably greater than one. Iron, alloys based on iron, nickel–iron alloys and ferrites (a particular type of iron compound) are ferromagnetic materials. Paramagnetic materials have relative permeabilities slightly greater than one. While ferromagnetic materials become strongly magnetized when in a magnetic field, paramagnetic materials become only very weakly magnetized. Aluminium and chromium are examples of paramagnetic materials. Diamagnetic materials have a relative permeability slightly less than one. This means that such a material becomes weakly magnetized when placed in a magnetic field but in a direction which is opposite to that of the magnetic field. Copper and gold are examples of diamagnetic materials.

Self-assessment questions

27 Which type of magnetic material is iron – ferromagnetic, paramagnetic or diamagnetic?

28 Of the three types of magnetic material, ferromagnetic, paramagnetic and diamagnetic, which has the highest value of relative permeability?

Magnetization curves for ferromagnetic materials

The graph shown in Figure 92(a) shows how the flux density B in a ferromagnetic material depends on the magnetic field intensity H. This graph is known as a magnetization curve. It shows that as the magnetic field intensity is increased from zero, so the flux density increases. Because the initial slope of the graph is not constant but increasing, this means that the value of $B–H$ is not constant, i.e. the permeability is not constant. The $B–H$ value gradually increases until it reaches a maximum, when the slope of the $B–H$ graph is a maximum. Then it decreases. This means that the permeability increases from its initial value to a maximum and then continually decreases. Figure 92(b) shows these permeability changes. The material is said to be magnetically saturated when the flux density becomes virtually constant and almost independent of the magnetic field intensity.

The above can be illustrated in a simple application with reference to an electromagnet, i.e. a coil wrapped round a length of ferromagnetic material. If the current through the coil is increased, then the m.m.f. is increased and so the magnetic field intensity (H = m.m.f./L) is increased. Initially an increase in the current leads to an increase in the flux density. However, this flux density only increases up to a particular value of the current, and beyond that value the flux density is almost constant and barely affected by further increases in current. This value of current is the one at which saturation first occurs. Thus as the strength of the electromagnet is related to the flux density then there is a current value beyond which it is pointless going to higher values as no increase in strength will occur.

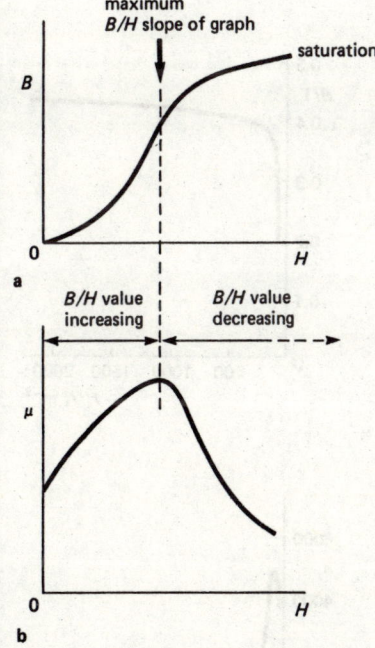

Figure 92 (a) *A magnetization curve*
(b) *The corresponding permeability-H graph*

Figure 93   (a) *Cast iron*
(b) *Mild steel*
(c) *Silicon-iron*
(d) *A ferrite*

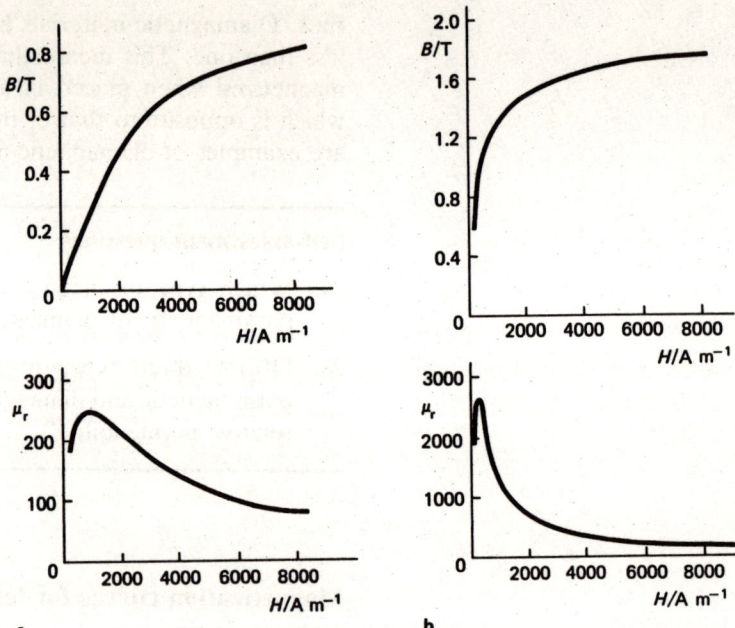

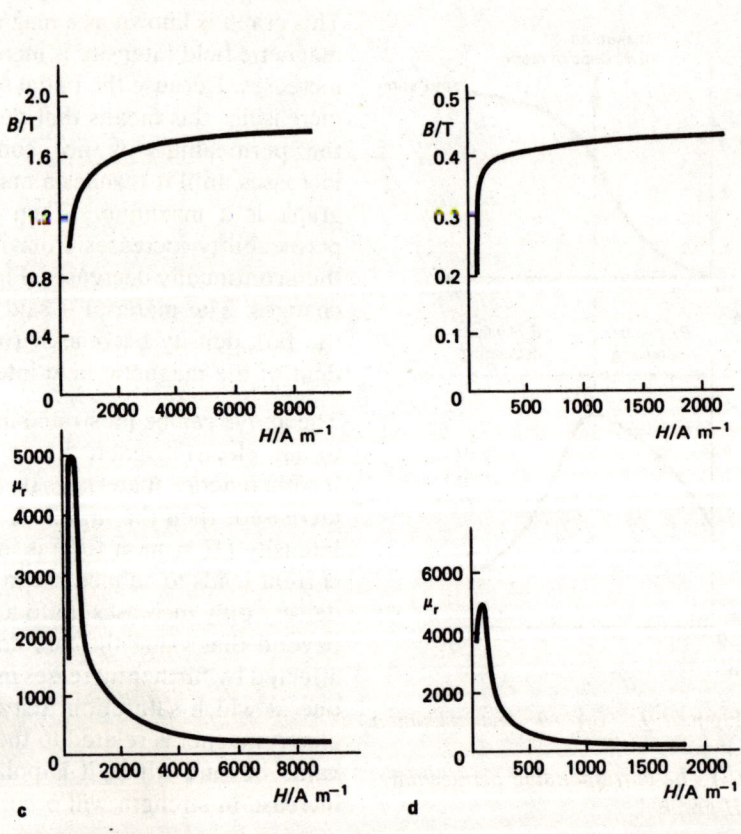

Figure 93 shows magnetization curves for a number of materials and their corresponding permeability–magnetic field intensity graphs. As will be immediately apparent, the relative permeability of a ferromagnetic material is far from constant. This means that in problems involving the relative permeability for 'real' materials the *B–H* graph has to be used to determine flux density values from magnetic field intensities, rather than just taking a value for the relative permeability.

Example 12

Using the *B–H* graph for cast iron given in Figure 93(a), what are the following values at a magnetic field intensity of 2000 A/m: (a) the flux density (b) the *B/H* value (c) the relative permeability?

(a)   The graph seems to indicate a flux density of about 0.5 T.

(b)   Thus $\dfrac{B}{H} = \dfrac{0.5}{2000}$ $\dfrac{\text{T}}{\text{A/m}}$

$$= 2.5 \times 10^{-4}\ \text{T m A}^{-1}$$
The above value is the absolute permeability.

(c) $\mu_r = \dfrac{\mu}{\mu_0} = \dfrac{2.5 \times 10^{-4}}{4\pi \times 10^{-7}}$ $\dfrac{\text{T m A}^{-1}}{\text{T m A}^{-1}}$

$$= 199$$

Example 13

Using the *B–H* graph for the ferrite given in Figure 93(d), what is the relative permeability at a magnetic field intensity of 250 A/m?

At $H = 250$ A/m the flux density is 0.4 T. Thus we have:

$$\mu = \frac{B}{H}$$

$$= \frac{0.4}{250}$$ $\dfrac{\text{T}}{\text{A/m}}$

$$= 1.6 \times 10^{-3}\ \text{T A m}^{-1}$$

Thus: $\mu_r = \dfrac{\mu}{\mu_0}$

$$= \frac{1.6 \times 10^{-3}}{4\pi \times 10^{-7}}$$ $\dfrac{\text{T A m}^{-1}}{\text{T A m}^{-1}}$

$$= 1273$$

Example 14

A cast iron ring with a mean diameter of 150 mm is wound with 500 turns of wire. What is the flux density in the iron when the current through the turns is 500 mA? The graph of B against H for cast iron is given in Figure 93(a).

$$L = 150 \text{ mm} = 150 \times 10^{-3} \text{ m}, N = 500, I = 500 \text{ mA} = 500 \times 10^{-3} \text{ A}$$

$$H = \frac{\text{m.m.f.}}{L} = \frac{NI}{L}$$

$$= \frac{500 \times 500 \times 10^{-3}}{\pi \times 150 \times 10^{-3}} \qquad \frac{\text{A}}{\text{m}}$$

$$= 541 \text{ A/m}$$

Using the graph, this magnetic field intensity indicates a flux density of about 0.18 T.

Example 15

A magnetic circuit has a mean length of 0.20 m of mild steel and an air gap of 2 mm. The cross-sectional area of the flux path is constant at 4 cm². What m.m.f. will be needed to produce a flux of 0.5 mWb in the air gap? The graph of B against H for mild steel is given in Figure 93(b).

For the mild steel: $L = 0.20$ m, $A = 4$ cm² $= 4 \times 10^{-4}$ m²,
$\Phi = 0.5$ mWb $= 0.5 \times 10^{-3}$ Wb

Note that because the circuit can be assumed to be series, the flux is constant at all parts of it, i.e. the flux in the mild steel is the same as that in the air gap.

Hence flux density in mild steel $B = \dfrac{\Phi}{A}$

$$= \frac{0.5 \times 10^{-3}}{4 \times 10^{-4}} \qquad \frac{\text{Wb}}{\text{m}^2}$$

$$= 1.25 \text{ T}$$

For the graph given in Figure 93(b), this flux density corresponds to a magnetic field intensity of about 800 A/m.

Hence the m.m.f. across the mild steel is given by:

$$H = \frac{\text{m.m.f.}}{L}$$

$$\text{m.m.f.} = H \times L$$
$$= 800 \times 0.20 \qquad \text{A/m} \times \text{m}$$
$$= 160 \text{ A}$$

For the air: $L = 2$ mm $= 2 \times 10^{-3}$ m, $A = 4$ cm^2 $= 4 \times 10^{-4}$ m^2, $\Phi = 0.5$ mWb $= 0.5 \times 10^{-3}$ Wb. We may also assume that for air $\mu_r = 1$ and is not affected by the value of B or H.

As before, $B = 1.25$ T

$$B = \mu H$$

Hence: $H = \dfrac{B}{\mu_r \mu_0}$

$$= \frac{1.25}{1 \times 4\pi \times 10^{-7}} \qquad \qquad \frac{\text{T}}{\text{T m A}^{-1}}$$

$$= 9.95 \times 10^5 \text{ A/m}$$

Hence the m.m.f. across the air gap is given by:

$$H = \frac{\text{m.m.f.}}{L}$$

m.m.f. $= H \times L$

$\qquad = 9.95 \times 10^5 \times 2 \times 10^{-3}$ $\qquad \qquad$ A/m $\times$ m

$\qquad = 1.99 \times 10^3$ A

The total m.m.f. required in the circuit is the sum of the m.m.f. across the mild steel and the m.m.f. across the air gap. It is just like adding the two potential differences across two resistors in series in an electrical circuit.

total m.m.f. $= 160 + 1.99 \times 10^3$ $\qquad \qquad$ A

$\qquad \qquad \quad = 2.15 \times 10^3$ A

Note that an alternative way of calculating the m.m.f. across the air gap would have been to calculate the reluctance and then use m.m.f. $= \Phi S$ to obtain the m.m.f.

Self-assessment questions

29 Using the B–H graph for silicon-iron given in Figure 93(c), what are the following values at a magnetic field intensity of 2000 A/m?
(a) The flux density
(b) The B–H value
(c) The permeability
(d) The relative permeability

30 Repeat Question 29 for a magnetic field intensity of 5000 A/m.

31 A mild steel ring with a mean diameter of 200 mm is wound with 1000 turns of wire. What is the flux density in the mild steel when the current through the turns is 600 mA. The B–H graph for mild steel is given in Figure 93(b).

32 A cast iron ring with a mean diameter of 120 mm is wound with 600 turns of wire. What should be the current through these turns if a flux density of 0.6 T is to be produced in the iron? The *B–H* graph for cast iron is given in Figure 93(a).

33 A magnetic circuit has a mean length of 0.25 m of mild steel and an air gap of 0.5 mm. The cross-sectional area of the flux path is constant at 6 cm². What m.m.f. will be needed to produce a flux of 0.6 mWb in the steel? The *B–H* graph for mild steel is given in Figure 93(b).

34 A magnetic circuit consists of 0.20 m of mild steel in series with 0.10 m of cast iron. The cross-sectional area of the circuit is constant at 4 cm².
(a) If the flux in the cast iron is 0.3 mWb, what is the flux in the mild steel?
(b) A coil of 400 turns is wound round the mild steel. What current through the coil is needed to produce the above flux?
The *B–H* graphs for mild steel and cast iron are given in Figure 93.

35 Explain what is meant by (a) the magnetization curve and (b) magnetic saturation.

Hysteresis

If a coil of wire is wrapped round an initially unmagnetized sample of a ferromagnetic material and the current gradually increased, we have the situation of a gradually increasing magnetic field intensity (remember $H = NI/L$, so H is proportional to the current I). The flux density B in the material increases. The result is the *B–H* graph illustrated in Figures 92 or 93. No consideration has been given to what happens if the current is then reduced or perhaps reversed in direction. In many situations in which magnetic materials are used the current is alternating current and so an alternating magnetic field intensity is being applied to the material. Figure 94 shows the type of *B–H* graph which is useful in such situations. It is known as the *B–H* loop.

Starting at zero magnetic field intensity, i.e. point P, the magnetic field intensity is gradually increased and the *B–H* graph line P to Q is produced. This is the magnetization curve and is the same as that described in the previous section, i.e. Figure 92. At Q we have magnetic saturation.

The magnetic field intensity is then gradually reduced to zero. The

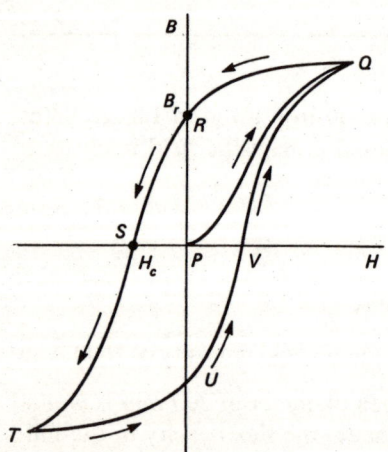

Figure 94 B–H *loop*

graph for this is from Q to R. When the magnetic field intensity is zero, i.e. point R, the material still has magnetic flux density. It is magnetized. If the material was then removed from the magnetizing coil we would have a permanent magnet. The flux density at this zero magnetic field intensity is called the remanent flux density B_r.

If the magnetic field intensity is reversed, by reversing the direction of the current in the magnetizing coil, the graph goes from R to S to T. Part of the way through this reversal, at point S, there is a magnetic field intensity but no flux density in the material. This is the value of the magnetic field intensity that is needed to demagnetize the material. The value of the magnetic field intensity at this condition is known as the coercive force, H_C. The term coercivity is used for the value of the coercive force when the initial magnetizing force has been sufficient to produce saturation.

At point T the material is magnetically saturated. If the magnetic field intensity is reduced to zero the graph goes from T to U. Again with zero magnetic field intensity there is still a magnetic flux intensity in the material, the remanent flux density.

If the magnetic field intensity is again reversed, the graph goes from U to V and back to Q. At point V, although there is a magnetic field intensity, there is no flux, as this is the coercive force situation.

If an alternating magnetic field intensity is used, with the peak to peak values being those corresponding to Q and T, then the flux density in the material follows the loop shown in the graph, repeatedly going round the loop. Figure 95 shows the effect of different values of peak to peak magnetic field intensities.

The loop represents an energy loss, known as the hysteresis loss. The greater the area enclosed by the loop the greater the energy loss. There is an energy loss every time the material is taken through the complete cycle represented by the hysteresis loop. Thus if alternating current is providing the alternating magnetic field intensity H, then every cycle of this current represents once round the loop and the energy loss given by the area enclosed. Thus if the alternating current has a frequency f, i.e. there are f complete cycles of current every second, then the energy loss per second is f times the area of the loop. The energy lost per second is the power loss, so the power loss is f times the area of the loop.

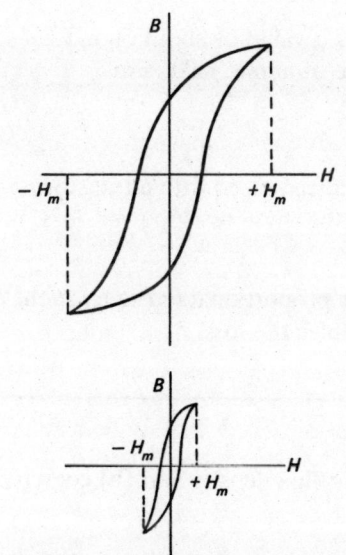

Figure 95 *The magnetic field intensity is oscillating between* $+H_m$ *and* $-H_m$, *the upper graph indicating a bigger value of peak to peak values than the lower graph*

Thus if the energy loss per cycle, i.e. the energy indicated by the area of the loop, is 500 J then the power loss with an alternating current of 50 Hz is $50 \times 500 = 25\,000$ J. If the frequency was 100 Hz, then the power loss would be $100 \times 500 = 50\,000$ J.

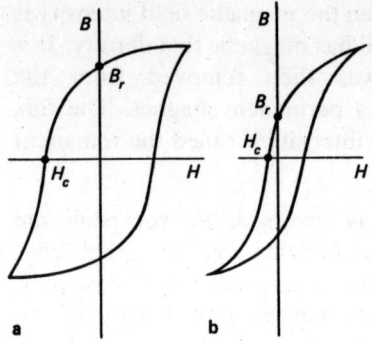

Figure 96   (a) *Type of material used for a permanent magnet*
(b) *Type of material used for an electromagnet*

Figure 96 shows typical hysteresis loops for the types of materials that could be used for a permanent magnet (see Figure 96(a)) and an electromagnet (see Figure 96(b)). The permanent magnet material has to be one that retains its magnetism when the magnetizing field is removed. It thus requires a high value of remanent flux density B_r. It also needs to be difficult to demagnetize and so needs a high value of coercive force H_C. The electromagnet, however, requires a material that does not retain its magnetism when the magnetizing field is removed and is also easy to demagnetize. This material thus needs a small remanent flux density B_r and a small coercive force H_C. The type of material shown in Figure 96(a), i.e. the permanent magnet material, is known as a hard magnetic material. The type of material shown by Figure 96(b), i.e. the electromagnet material, is known as a soft magnetic material.

In some situations, such as transformer cores, the important requirement of the material is that it be low loss. Special alloys have been developed for this purpose and these materials have hysteresis loops looking like that shown in Figure 97.

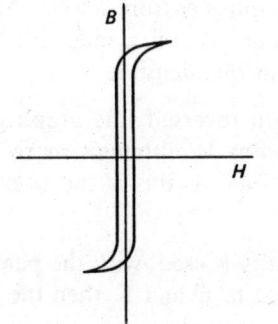

Figure 97   *A low loss material*

Example 16
For the hysteresis loop illustrated in Figure 98, what are (a) the remanent flux density and (b) the coercive force?

(a)   The remanent flux density has a value of about 50 mT.
(b)   The coercive force has a value of about 3000 A/m.

Example 17
If the frequency of the alternating current used with a transformer is doubled, by what factor will the hysteresis power loss be changed?

The hysteresis power loss is directly proportional to the frequency and so doubling the frequency doubles the loss.

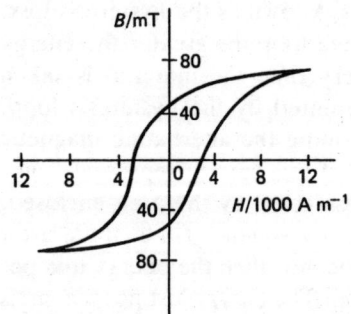

Figure 98

Self-assessment questions

36   Explain the terms (a) remanent flux density and (b) coercive force.

37   How does the hysteresis loop for a soft magnetic material differ from that for a hard magnetic material?

38   How does the hysteresis loop for a low loss material differ from that for a high loss material?

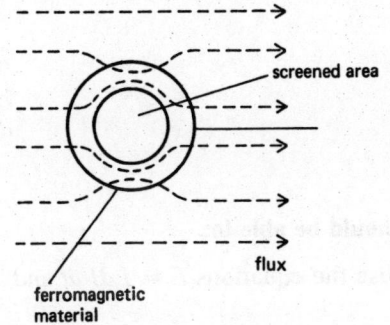

Figure 99   *Magnetic screening*

Magnetic screening

There are some circumstances where it is required that a component or some instrument is screened from magnetic fields. An example of this is the magnetic recording head in a tape recorder. Magnetic flux tends to be found to be concentrated in those parts where the reluctance is smallest. Thus if a component is surrounded by a low reluctance magnetic material, the magnetic flux will tend to concentrate in that material rather than pass through the space with the higher reluctance. Figure 99 shows such a form of screening where the low reluctance path is provided by a ferromagnetic material and the high reluctance path by air.

Self-assessment question

39 Why does surrounding a component with ferromagnetic material screen it from magnetic fields?

5 Inductance

After reading this chapter you should be able to:

1 Explain self inductance and use the equations $E = LdI/dt$ and $L = Nd\Phi/dt$.
2 Derive and use the equation for the self inductance of a coil.
3 Explain arcing in inductive circuits.
4 Derive and use the relationship for the energy stored in an inductor.
5 Explain mutual inductance and use the equation $E = MdI/dt$.
6 Explain the basic principles of the transformer and solve problems involving the turns ratio.

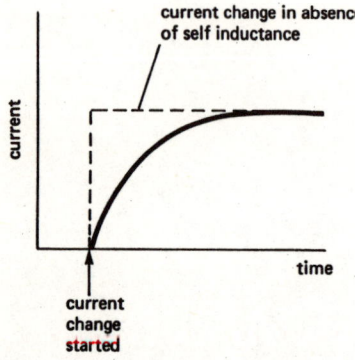

Figure 100 *The effect of self induct-ance on an increase in current*

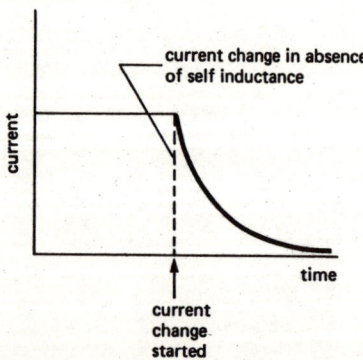

Figure 101 *The effect of self induct-ance on a decrease in current*

Self inductance

When a current flows through a coil it gives rise to a magnetic field, i.e. magnetic flux links the coil. If the current through the coil changes then the magnetic field changes and so the flux linked by the coil changes. If there is a change in the magnetic flux linked by a coil then electromagnetic induction occurs and an e.m.f. is induced in the coil. This e.m.f. is in such a direction as to oppose the change producing it. Thus if the current through the coil is being increased, the induced e.m.f. is in such a direction as to produce a current which slows down the growth in the current (see Figure 100). If the current through the coil is being decreased, the induced e.m.f. is in such a direction as to produce a current which slows down the rate of decrease, i.e. the current takes longer to become reduced (see Figure 101). This effect is known as self inductance, since it is caused by changes in the current in the coil itself.

A circuit is said to have a self inductance of one henry (H) if, when the current in the circuit changes at the rate of one ampere per second, an e.m.f. of one volt is induced in it.

This can be expressed in an equation:

self induced e.m.f. = L × rate of change of current

$$E = L \frac{dI}{dt}$$

The symbol L is used for self inductance, dI/dt is the notation used for rate of change of current with time. The above equation is sometimes written with a minus sign, i.e. $E = - LdI/dt$, to indicate that the direction of the self induced e.m.f. is in such a direction as to oppose the rate of change of current producing it.

For a coil with N turns and flux changing at the rate of $d\Phi/dt$, the induced e.m.f., according to Neumann's equation (see Chapter 3), is given by:

$$E = N \frac{d\Phi}{dt}$$

However, if this e.m.f. is produced as a result of self inductance, we have:

$$E = L \frac{dI}{dt}$$

Hence: $\quad L \dfrac{dI}{dt} = N \dfrac{d\Phi}{dt}$

and so: $\quad \boxed{L = N \dfrac{d\Phi}{dI}}$

Expressing the above equation in words:

$$L = N \times \frac{\text{change in flux linked}}{\text{change in current}}$$

Example 1
What is the e.m.f. induced in a circuit of inductance 0.4 H when the current changes at 10 A/s?

$L = 0.4$ H, $dI/dt = 10$ A/s

$$E = L \frac{dI}{dt}$$

$$\begin{aligned} &= 0.4 \times 10 && \text{H} \times \text{A/s} \\ &= 4.0 \text{ V} \end{aligned}$$

Example 2
Calculate the inductance of a coil if an e.m.f. of 50 V is induced in it when the current changes at the rate of 100 A/s?

$E = 50$ V, $dI/dt = 100$ A/s

$$E = L \frac{dI}{dt}$$

Hence: $L = \dfrac{E}{(dI/dt)}$

$$= \frac{50}{100} \qquad\qquad \frac{V}{A/s}$$

$$= 0.5 \text{ H}$$

Example 3

Calculate the average e.m.f. induced in a coil of inductance 0.2 H when the current through it changes from 1 A to 2.5 A in 0.1 s.

$$\text{Average rate of change of current} = \frac{2.5 - 1.0}{0.1} \qquad \frac{A}{s}$$

$$= 15 \text{ A/s}$$

Thus we have $E = 0.2$ H, average $dI/dt = 15$ A/s

$$E = L \frac{dI}{dt}$$

$$= 0.2 \times 15 \qquad\qquad\qquad \text{H} \times \text{A/s}$$
$$= 3 \text{ V}$$

This is the average induced e.m.f. because we only know the average rate of change of current.

Example 4

When the current changes from 2 to 4 A in a coil of 1000 turns, the flux linked by the coil changes by 40 μWb. What is the inductance of the coil? (Assume the changes occur at a constant rate.)

Change in current $= 4 - 2 = 2$ A, change in flux $= 40\ \mu\text{Wb} = 40 \times 10^{-6}$ Wb, $N = 1000$

$$L = N \frac{d\Phi}{dI}$$

$$= 1000 \times \frac{40 \times 10^{-6}}{2} \qquad\qquad \frac{Wb}{A}$$

$$= 2 \times 10^{-2} \text{ H}$$

Self-assessment questions

1 Explain what is meant by self inductance.

2 Define the henry.

3 What is the e.m.f. induced in a circuit of inductance 500 mH when the current changes at 20 A/s?

4 Calculate the inductance of a coil if an e.m.f. of 30 V is induced in it when the current changes at the rate of 60 A/s.

5 When a current of 4 A flowing through a coil is reversed in 0.05 s, what is the induced e.m.f. if the coil has an inductance of 0.2 H?

6 Calculate the average e.m.f. induced in a coil of inductance 0.5 H when the current through it changes from 0.5 A to 2 A in 0.2 s.

7 When the current in a coil of 800 turns changes from zero to 2 A the flux linked by the coil changes by 50 μWb. What is the inductance of the coil? (Assume the changes occur at a constant rate.)

The inductance of a coil

Consider a coil of N turns wound on a core of magnetic length l and cross-sectional area A. When a current I flows through the coil the flux linked by each turn is Φ. If the flux is proportional to the current (the case in an air-cored coil), then as a change in flux of $(\Phi - 0)$ is produced when the current changes from 0 to I we have:

$$\frac{d\Phi}{dI} = \frac{\Phi}{I}$$

Hence as: $L = N \dfrac{d\Phi}{dI}$

we have: $\boxed{L = \dfrac{N\Phi}{I}}$

But m.m.f. $= \Phi S$, where the m.m.f. $= NI$ and $S = l/(\mu A)$, hence:

$$NI = \Phi \times \frac{l}{\mu A}$$

and so: $\Phi = \dfrac{NI\,\mu A}{l}$

Hence: $L = \dfrac{N\Phi}{I} = \dfrac{N}{I} \times \dfrac{NI\,\mu A}{l}$

$$\boxed{L = \dfrac{N^2\,\mu A}{l}}$$

The inductance is proportional to the square of the number of turns, proportional to the area A and inversely proportional to the length of the coil. Because of the μ factor, an iron-cored coil will have a higher inductance than an air-cored coil.

Example 5
When a current of 2 A flows in a 200 turn air-cored coil, a flux of 0.2 mWb is produced. What is the inductance of the coil?

$I = 2$ A, $N = 200$, $\Phi = 0.2$ mWb $= 0.2 \times 10^{-3}$ Wb

$L = \dfrac{N\Phi}{I}$

$= \dfrac{200 \times 0.2 \times 10^{-3}}{2}$ $\qquad\qquad \dfrac{\text{Wb}}{\text{A}}$

$= 0.02$ H

Example 6
Calculate the inductance of a coil with an air core and 600 turns occupying a length of 0.2 m and a cross-sectional area of 6 cm².

$N = 600$, $l = 0.2$ m, $A = 6$ cm² $= 6 \times 10^{-4}$ m², $\mu_0 = 4\pi \times 10^{-7}$ T m A^{-1}

$L = \dfrac{N^2\,\mu A}{l}$

$= \dfrac{600^2 \times 4\pi \times 10^{-7} \times 6 \times 10^{-4}}{0.2}$ $\qquad \dfrac{\text{T m A}^{-1} \times \text{A}}{\text{m}}$

$= 1.36 \times 10^{-3}$ H

Example 7
If the number of turns of the coil in the previous question are doubled, without any change in length or cross-sectional area, how is the inductance changed?

$L \propto N^2$

Thus doubling N means an increase in L by a factor of 2^2, i.e. 4. The inductance is thus quadrupled.

Self-assessment questions

8 A flux of 2 mWb is produced in an air-cored coil of 2000 turns when a current of 6 A flows through it. What is the inductance of the coil?

9 An air-cored coil has an inductance of 0.03 H. If the coil has 500 turns, what is the flux produced by a current of 1.5 A?

10 Calculate the inductance of an air-cored coil with 1000 turns, a length of 0.15 m and a cross-sectional area of 4 cm^2.

11 A spring is used as a coil. If the spring is stretched to double its length, the area decreasing to one quarter of its initial value, how is the inductance of the coil changed?

Arcing in inductive circuits

The induced e.m.f. depends on the rate of change of current in an inductive circuit, the greater the rate of change of current the greater the induced e.m.f. Thus when an inductive circuit is switched off a very high voltage can be produced between the contacts of the switch. This can lead to a breakdown of the insulation and arcing across the contacts. This arcing can occur in domestic electrical circuits when, for example, a power circuit or a lighting circuit is switched off. In large installations or power stations, circuit breakers have to be used to avoid arcing when circuits are switched off.

Self-assessment questions

12 Explain why sparks can occur across the contacts of a switch when it is used to switch off some electrical circuit.

13 What is the average induced e.m.f. in a circuit which has an inductance of 0.2 H when a current of 2 A is switched off and falls to zero in 0.01 s?

14 An electrical switch is replaced by one which switches off a current at a rate ten times faster than before. What problems might this lead to?

Energy stored in an inductor

When the current through an inductor is switched on, the magnetic field grows and reaches a steady value when the current reaches its steady value. The energy for creating this magnetic field comes from the d.c. power source used to produce the current. When the current is switched off, there is no power source in the circuit but a current continues for some period of time. The energy which makes this current continue after the power source is switched off comes from the magnetic field of the inductor.

Consider the current through an inductor increasing from zero to I during some time interval t.

average induced e.m.f. = L × rate of change of current

The average rate of change of current is I/t, hence:

$$\text{average induced e.m.f.} = \frac{LI}{t}$$

The average energy consumed by the circuit during this time is:

average energy = average power × time

Power = IV, so the average power is the product of the average current and the average voltage. The average current is $\frac{1}{2}I$, thus:

$$\text{average energy} = \frac{1}{2}I \times \frac{LI}{t} \times t$$
$$= \frac{1}{2}LI^2$$

Example 8

Calculate the energy stored in a coil which has an inductance of 2 H when a current of 4 A passes through it.

$L = 2$ H, $I = 4$ A
$$\text{energy} = \frac{1}{2}LI^2$$
$$= \frac{1}{2} \times 2 \times 4^2 \qquad\qquad\qquad \text{H} \times \text{A}^2$$
$$= 16 \text{ J}$$

Self-assessment questions

15 Calculate the energy stored in a coil with an inductance of 0.5 H when a current of 3 A passes through it.

16 Calculate the energy stored in an inductor of 10 H when a current of 2 A flows through it.

17 Calculate the energy stored in a coil of 400 turns on an iron core, when the current through it is 2 A and this results in a flux of 3 mWb.

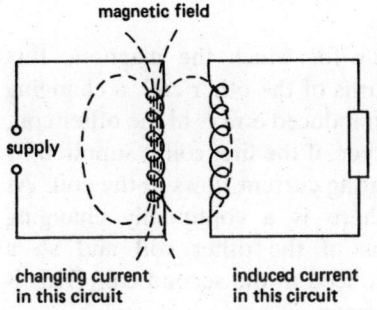

magnetic field

supply

changing current
in this circuit

induced current
in this circuit

Figure 102 *Mutual inductance*

Mutual inductance

When a current flows in a coil a magnetic field is produced. If the current is changing, a changing magnetic field is produced. If nearby there is another coil then the changing magnetic field can induce an e.m.f. and hence a current in this coil. Figure 102 illustrates this effect.

The two coils in Figure 102 can be considered to be coupled by magnetic flux, i.e. the magnetic flux produced in one coil links the turns in the other coil. We use the term mutual inductance to describe this effect, i.e. that a change of current in one circuit produces an e.m.f. in the other circuit.

Two coupled circuits have a mutual inductance of one henry (H) if, when the current changes in one circuit at the rate of one ampere per second, an e.m.f. of one volt is induced in the other circuit.

This can be expressed by the equation:

induced e.m.f. in one circuit = mutual inductance × rate of change of current in the other circuit

$$E_1 = M \frac{dI_2}{dt}$$

where M is the mutual inductance, E_1 the induced e.m.f. in circuit number 1, and dI_2/dt the rate of change of current in circuit number 2.

Example 9

If the mutual inductance between a pair of coils is 200 mH, what is the e.m.f. induced in one of the coils when the rate of change of current in the other coil is 10 A/s?

$M = 200 \text{ mH} = 200 \times 10^{-3} \text{ H}$, $dI_2/dt = 10$ A/s

$E_1 = M \frac{dI_2}{dt}$

$= 200 \times 10^{-3} \times 10$ H × A/s

$= 2.0$ V

Self-assessment questions

18 Explain what is meant by mutual inductance.

19 If the mutual inductance between a pair of coils is 100 mH, what is the e.m.f. induced in one coil when the rate of change of current in the other coil is 20 A/s?

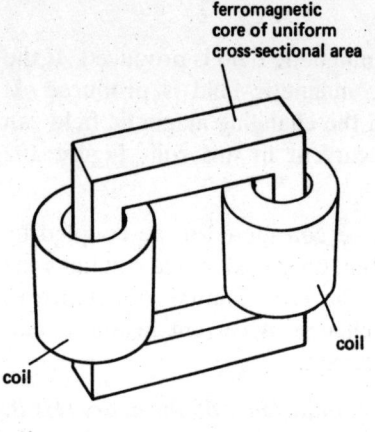

ferromagnetic core of uniform cross-sectional area

coil

coil

a

The transformer

For two coupled coils, i.e. coils for which the magnetic flux produced by one coil links the turns of the other coil, a changing current in one coil gives rise to an induced e.m.f. in the other coil. This is mutual inductance. However, if the first coil is supplied by an alternating voltage, an alternating current flows in the coil. As this is continually changing, there is a continually changing magnetic flux linking the turns of the other coil and so a continually changing e.m.f. is induced in the second coil. This is the basic principle of the transformer.

A transformer can be considered to be an electromagnetic device which 'transforms' an alternating voltage in one circuit into an alternating voltage in another circuit.

Figure 103 shows one form of a transformer. It consists of one coil, known as the primary, to which the alternating voltage input is connected and another coil, known as the secondary, in which the e.m.f. is induced. The coils are both wound round the same magnetic core, so virtually all the flux produced by the primary coil links the secondary coil. In this particular form of transformer, known as a core-type transformer, the magnetic circuit is of ferromagnetic material of uniform cross-sectional area.

Figure 104 shows another form of transformer, known as the shell-type. In this transformer both the primary and the secondary coils are wound on the central limb of the magnetic circuit. The central limb has a cross-sectional area twice that of the two outer limbs. This is because the central limb carries twice the flux of the two outer limbs. The circuit symbol and principle of operation is the same as the core-type transformer.

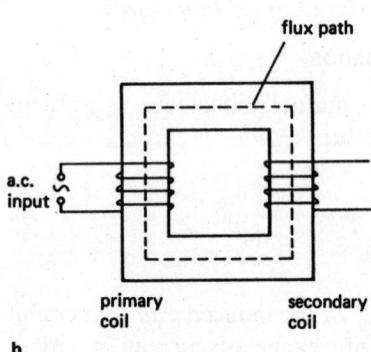

flux path

a.c. input

primary coil

secondary coil

b

c

Figure 103   *A core-type transformer*
(a) *The construction*
(b) *The magnetic circuit*
(c) *Circuit symbol*

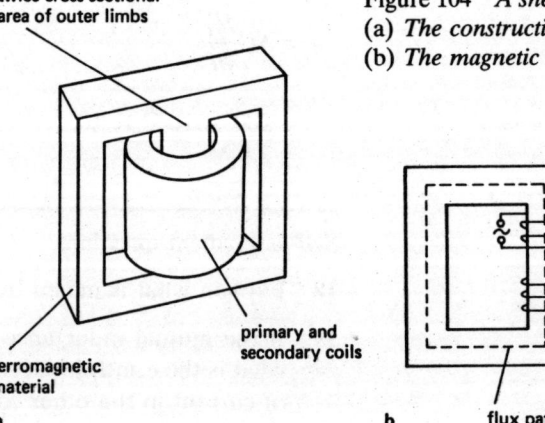

twice cross-sectional area of outer limbs

ferromagnetic material

primary and secondary coils

a

Figure 104   *A shell-type transformer*
(a) *The construction*
(b) *The magnetic circuit*

primary

secondary

flux path

b

In an ideal transformer, i.e. a transformer where all the flux produced by one of the transformer coils links with the windings of the other coil, the resistance of the primary circuit is zero and the secondary is an open circuit. In the case of a primary coil, if the current change produces a rate of flux change of $d\Phi/dt$, then the e.m.f. induced in the primary coil by such a flux change is, using Neumann's equation:

$$\text{induced e.m.f.} = N_p\frac{d\Phi}{dt}$$

where N_p is the number of turns in the primary coil. But this flux change also links the secondary coil, the e.m.f. induced in the secondary coil E_s is:

$$E_s = N_s\frac{d\Phi}{dt}$$

where N_s is the number of turns in the secondary coil.

If E_p is the input e.m.f. to the primary coil, the net e.m.f. in the primary circuit must be:

$$E_p - N_p\frac{d\Phi}{dt}$$

The induced e.m.f. opposes the change producing it and so is in the opposite direction to the input e.m.f. This net e.m.f. must produce a current I across the circuit resistance R, i.e.:

$$E_p - N_p\frac{d\Phi}{dt} = IR$$

But because we have assumed this is an ideal transformer, the primary circuit resistance has been taken to be zero. Thus:

$$E_p - N_p\frac{d\Phi}{dt} = 0$$

$$\text{and:} \quad E_p = N_p\frac{d\Phi}{dt}$$

Hence the ratio of the primary input e.m.f. to the secondary e.m.f. is

$$\frac{E_p}{E_s} = \frac{N_p\dfrac{d\Phi}{dt}}{N_s\dfrac{d\Phi}{dt}}$$

$$\boxed{\frac{E_p}{E_s} = \frac{N_p}{N_s}}$$

N_p/N_s is known as the turns ratio. If N_p is greater than N_s then the primary e.m.f. is greater than the secondary e.m.f., the transformer is then said to have a step-down voltage ratio. If N_p is less than N_s then the primary e.m.f. is less than the secondary e.m.f. and the transformer is said to have step-up voltage ratio.

It may seem that with a step-up voltage ratio you get 'something-for-nothing', but consider the power input and output for the transformer when there is a load in the secondary circuit. If we neglect any power losses (the transformer efficiency is very high anyway), then the input power in the primary must equal the power output in the secondary. To a reasonable approximation we can state that:

$$\text{input power} = I_p E_p$$
$$\text{and: } \text{output power} = I_s E_s$$
$$\text{and so: } I_p E_p = I_s E_s$$

Thus:
$$\frac{E_p}{E_s} = \frac{I_s}{I_p}$$

and so:
$$\boxed{\frac{E_p}{E_s} = \frac{I_s}{I_p} = \frac{N_p}{N_s}}$$

When the transformer steps up the voltage it steps down the current. When it steps down the voltage it steps up the current.

An important point to remember is that in the above example the input and the output of the transformer are both alternating. The frequencies of the input voltage and current are the same as the frequencies of the output voltage and current.

Example 10
What is the turns ratio required for a transformer used to step down the 240 V mains alternating current to a 12 V alternating voltage?

$E_p = 240$ V, $E_s = 12$ V

$$\frac{E_p}{E_s} = \frac{N_p}{N_s} = \text{turns ratio}$$

Hence: turns ratio $= \dfrac{240}{12}$ $\dfrac{\text{V}}{\text{V}}$

$$= 20$$

Example 11
A transformer has a step-up voltage ratio of 8:1. If the primary voltage is 24 V, what is the secondary voltage?

The term 'step-up' means that the secondary voltage is higher than the primary voltage, by the voltage ratio of 8:1. This means that the secondary voltage must be 8 times the primary voltage, i.e. $8 \times 24 = 192$ V.

Example 12
An ideal transformer has a primary coil with 200 turns and a secondary coil with 40 turns. How are (a) the secondary and primary voltages related and (b) the secondary and primary currents related?

$N_p = 200$, $N_s = 40$

(a) $\dfrac{E_p}{E_s} = \dfrac{N_p}{N_s}$

$= \dfrac{200}{40}$

$= 5$

The primary voltage is 5 times the secondary voltage.

(b) $\dfrac{I_s}{I_p} = \dfrac{N_p}{N_s}$

$= \dfrac{200}{40}$

$= 5$

The secondary current is 5 times the primary current.

Self-assessment questions

20 What is the turns ratio required for a transformer used to step up an alternating voltage of 6 V to one of 60 V?

21 A transformer has a step-up voltage ratio of 10:1. If the primary voltage is 240 V, what is the secondary voltage?

22 A transformer has a step-down voltage ratio of 5:1. If the primary coil has 400 turns, what is the number of turns in the secondary coil?

23 An ideal transformer has 600 turns of wire in the primary and 50 turns in the secondary. If the input voltage is 12 V and the current 2 A, what is (a) the output voltage (b) the output current?

6 Alternating voltages and currents

After reading this chapter you should be able to:

1 Distinguish between alternating and unidirectional waveforms.
2 Explain and use the terms peak value, cycle, periodic time, frequency and instantaneous value.
3 Describe the sinusoidal waveform and use equations of the form $v = V_m\sin\theta$, $v = V_m\sin\omega t$ and $v = V_m\sin 2\pi ft$.
4 Determine average values using the mid-ordinate rule and, for sinusoidal waveforms, also by calculation.
5 Define the r.m.s. value and determine it using the mid-ordinate rule and, for sinusoidal waveforms, also by calculation.
6 Define and calculate form factors.
7 Explain what phasors are and how they can be added.
8 Explain what is meant by phase difference.
9 Explain the equation $v = V_m\sin(\omega t + \phi)$
10 Explain the principles of half wave and full wave rectification.

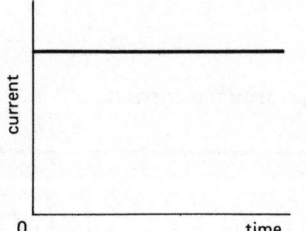

Figure 105 *A constant direct current*

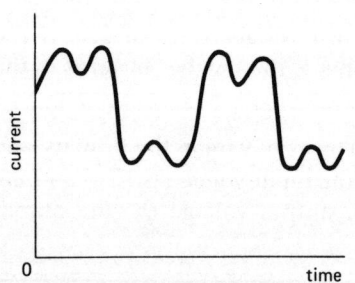

Figure 106 *A fluctuating direct current*

Waveforms

When a battery is connected to a resistive circuit a steady current is produced which always flows in the same direction round the circuit. Such a current is called a direct current. Figure 105 shows a graph of such a current with time. Direct currents do not have to be constant: they can vary with time, but they never change the direction in which they flow round the circuit. Figure 106 shows a fluctuating direct current and how it varies with time.

Similarly, direct voltages do not vary in direction in a circuit. They can be constant direct voltages, like the constant direct current in Figure 105, or fluctuating direct voltages like the fluctuating direct current in Figure 106.

An alternating current or an alternating voltage, on the other hand, does not have the same direction through the circuit, but its

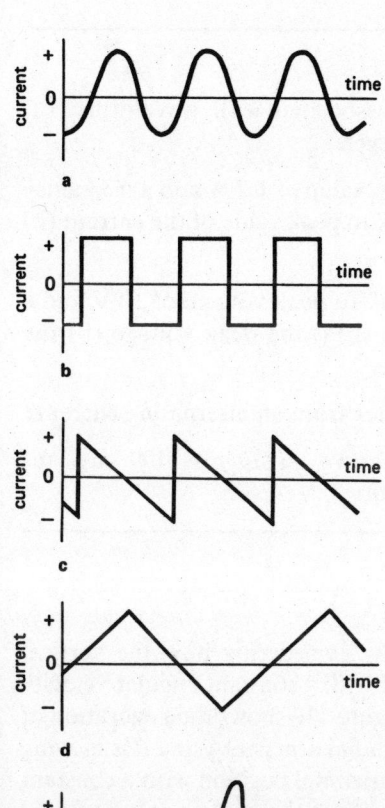

direction varies with time. Figure 107 shows some examples of alternating currents. The most common form is the sinusoidal waveform (see Figure 107(a)). The waveform is given this name because the graph is that of the sine function.

There are a number of terms used to describe a regular alternating current. Figure 108 illustrates these for the sinusoidal waveform.

The maximum size of a current or a potential difference is called the maximum or peak value or the amplitude of the waveform.

The peak to peak value is the vertical distance between the positive maximum and the negative maximum of the waveform. For a regular waveform this peak to peak value is twice the maximum or peak value.

One cycle of a waveform is one complete wave before the waveform repeats itself again. The time taken for one cycle is called the periodic time. The frequency is the number of cycles completed per second.

Thus if the periodic time is T then in one second the number of cycles possible is $1/T$. Hence the frequency f is:

$$f = \frac{1}{T}$$

When the time is in seconds the frequency is in hertz (Hz).

The term instantaneous current, or voltage, is used for the value of the current, or voltage at a particular instant. This value might be positive or negative, depending on which direction the current or voltage is at the instant concerned.

Figure 107 *Alternating currents*
(a) *Sinusoidal*
(b) *Rectangular*
(c) *Sawtooth*
(d) *Triangular*
(e) *Irregular*

Example 1
A sinusoidal alternating current has a peak value of 2 A and a periodic time of 0.02 s. What is (a) the peak to peak value (b) the frequency?

(a) peak to peak value = 2 × peak value

                             = 2 × 2   A

                             = 4 A

(b) $f = \dfrac{1}{T}$

                        $= \dfrac{1}{0.02}$ $\dfrac{1}{s}$

                        = 50 Hz

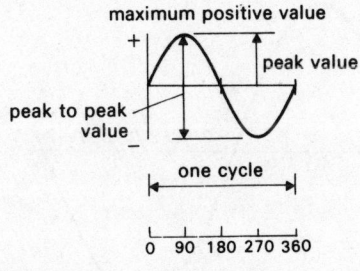

maximum positive value

peak value

peak to peak value

one cycle

0 90 180 270 360

Figure 108 *Alternating waveform terms*

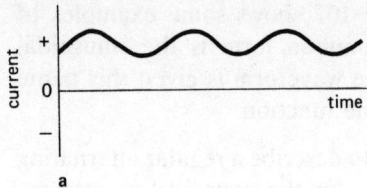

a

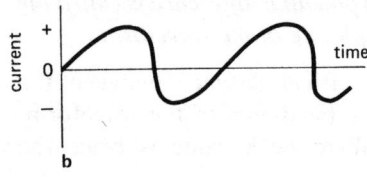

b

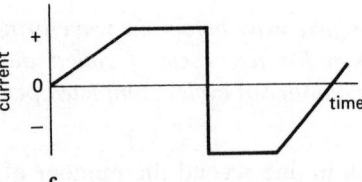

c

Figure 109

Self-assessment questions

1 Define the following terms associated with waveforms: (a) frequency (b) peak value (c) cycle.

2 A sinusoidal current has a peak value of 1.2 A and a frequency of 100 Hz. What is (a) the peak to peak value of the current (b) the periodic time?

3 A sinusoidal voltage has a peak to peak voltage of 10 V and a periodic time of 0.05 s. What is (a) the peak voltage (b) the frequency?

4 How does a direct current differ from an alternating current?

5 Which of the waveforms shown in Figure 109 are (a) unidirectional and (b) alternating?

The sine wave

A sine wave can be generated by considering how the vertical projection of a radial arm rotating with a constant angular velocity round a circle varies with time. Figure 110 shows the generation of a sine wave by this method. The radial arm is *OA* and it is moving round the circle from an initial horizontal position with a constant angular velocity. This means it covers the same angle in equal intervals of time. The vertical projection of this rotating radial arm when it has moved through 30° is *AB*. By measuring the vertical heights for different angles, a graph of vertical height against angle can be plotted. That is the graph known as the sine wave.

The maximum, or peak value, occurs when the radial arm has moved through 90° or 270°. The zero values occur when the radial arm is at 0° or has moved through 18(?) or 360°.

If we are concerned with, say, voltages, then the maximum or peak value V_m is the vertical height of the radial arm at 90°. But

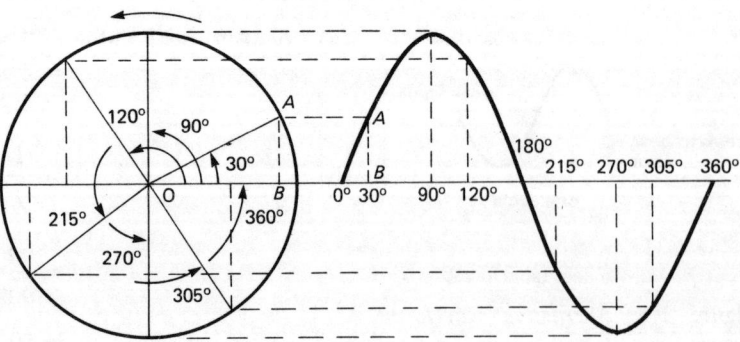

Figure 110 *Generating a sine wave*

this is the length of the radial arm. In the case of the radial arm at 30°, we have:

$$\frac{AB}{OA} = \sin 30°$$

But $OA = V_m$. v is the instantaneous value, in this case the value of the vertical projection AB at 30°, so:

$$\frac{v}{V_m} = \sin 30°$$
$$v = V_m \sin 30°$$

In general, we can write for any angle θ,

$$\boxed{v = V_m \sin \theta}$$

Note: The convention generally followed, and adopted in this book, is to use lower case (small) letters for instantaneous values and capital letters for values which do not depend on time.

In the above equation the angle has been assumed to be in degrees. Another form of angular measure is the radian. This is the measure that is more commonly used in alternating current theory.

The radian, generally abbreviated to rad, is the angle subtended at the centre of a circle by an arc with a length equal to the radius (see Figure 111).

As the circumference of a circle has a circumference of $2\pi R$, where R is the radius, then the number of radians in an angle equivalent to a rotation through a complete circle is:

$$\text{radians in a circle} = \frac{\text{circumference}}{\text{radius}}$$

$$= \frac{2\pi R}{R} = 2\pi$$

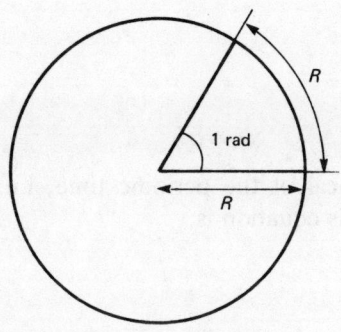

Figure 111   *One radian*

There are thus 2π radians involved in a rotation through a complete circle. In degree measure this is a rotation of 360°. Hence:

$$\boxed{2\pi \text{ rads} = 360°}$$

$$1 \text{ rad} = \frac{360°}{2\pi}$$

$$\boxed{1 \text{ rad} = 57.3°}$$

The following table shows some angles in radian and in degree measure:

0°	0 rad
90°	$\frac{1}{2}\pi$ rad
180°	π rad
270°	$1\frac{1}{2}\pi$ rad
360°	2π rad

Now to return to the equation $v = V_m \sin\theta$. If we have the radial arm rotating with a constant angular velocity ω (the unit of which is the rad/s), then in a time t the angle swept out by the arm is:

$$\theta = \omega t$$

where θ is in radians. Substituting this in the equation gives:

$$\boxed{v = V_m \sin\omega t}$$

If the frequency of the rotation is f then f complete rotations are completed every second. One complete rotation is 2π radians and so the angle swept out in one second is $2\pi f$ radians. But the angle swept out per second is the angular velocity ω, hence:

$$\boxed{\omega = 2\pi f}$$

and so we can write:

$$\boxed{v = V_m \sin 2\pi ft}$$

As the frequency f is the reciprocal of the periodic time, i.e. $f = 1/T$, another way of writing this equation is:

$$\boxed{v = V_m \sin\left(\frac{2\pi t}{T}\right)}$$

Example 2
Convert 30° into radians.

As 360° = 2π radians, we have:

$$1° = \frac{2\pi}{360} \text{ rad}$$

$$30° = \frac{2\pi}{360} \times 30 \qquad\qquad \text{rad}$$

$$= \frac{\pi}{12} \text{ rad}$$

Example 3

Calculate the instantaneous value of a sinusoidal voltage waveform which has a peak voltage of 6 V when $\theta = 20°$.

$v = V_m \sin \theta$
  $= 6 \sin 30°$
  $= 6 \times 0.5$
  $= 3$ V

Example 4

Calculate the instantaneous value of a sinusoidal voltage waveform which has a peak voltage of 4 V when $\theta = \frac{\pi}{4}$ rad.

As $360° = 2\pi$ radians we have:

$$\pi \text{ radians} = \frac{360°}{2}$$

$$\frac{\pi}{4} \text{ radians} = \frac{360°}{2} \times \frac{1}{4}$$

$$= 45°$$
$$v = V_m \sin \theta$$
$$v = 4 \sin 45°$$
$$= 4 \times 0.707$$
$$= 2.83 \text{ V}$$

Example 5

Calculate the instantaneous value of a sinusoidal voltage waveform which has a peak voltage of 5 V, a frequency of 50 Hz, and for which a time of 0.002 s has elapsed since the voltage was zero.

$v = V_m \sin 2\pi ft$
  $= 5 \sin (2\pi \times 50 \times 0.002)$
  $= 5 \sin 0.2\pi$

An angle of 0.2π radians is an angle of:

$$0.2\pi \text{ radians} = \frac{360°}{2} \times 0.2$$
$$= 36°$$

Hence $v = 5 \sin 36°$
      $= 5 \times 0.588$
      $= 2.94$ V

Example 6

Calculate the instantaneous value of a sinusoidal current waveform which has a peak current of 2 A, a frequency of 50 Hz, and for which a time of 0.003 s has elapsed since the current was zero.

$$i = I_m \sin 2\pi ft$$
$$= 2\sin (2\pi \times 50 \times 0.003)$$
$$= 2\sin 0.3\pi$$

An angle of 0.3π radians is an angle of:

$$0.3\pi \text{ radians} = \frac{360°}{2} \times 0.3$$
$$= 54°$$

Hence $i = 2\sin 54°$
$$= 2 \times 0.809$$
$$= 1.62 \text{ A}$$

Self-assessment questions

6   Convert 60° into radians.

7   Calculate the instantaneous values of the following sinusoidal voltage waveforms:
   (a)   A peak voltage of 6 V when $\theta = 45°$
   (b)   A peak voltage of 20 V when $\theta = 15°$
   (c)   A peak voltage of 12 V when $\theta = \frac{1}{3}\pi$
   (d)   A peak voltage of 6 V when $\theta = 0.4\pi$
   (e)   A peak voltage of 10 V, a frequency of 50 Hz, and for which a time of 0.003 s has elapsed since the voltage was zero
   (f)   A peak voltage of 4 V, a periodic time of 0.03 s and for which a time of 0.005 s has elapsed since the voltage was zero

8   Sketch the waveform of a sinusoidal voltage which has a peak voltage of 4 V and a frequency of 50 Hz. Clearly indicate the times taken from the start of the wave, at zero voltage, for the peak values and the zero values to occur.

Average values

If the positive part of the waveform of an alternating quantity is the same size and shape as the negative part, as in Figure 108 or 110, then for every positive value there is an equal sized negative value, so the average over an entire cycle is zero. For that reason we only consider the average value of an alternating waveform over one half-cycle.

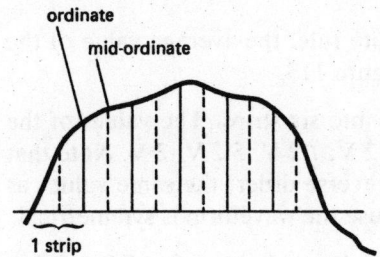

ordinate

mid-ordinate

1 strip

Figure 112 *Using the mid-ordinate rule with six strips*

One method of determining the average value is to use the mid-ordinate rule. The procedure to be used with this rule is as follows:

1 Divide the area under the graph for the half-cycle into a number of strips of equal width by ordinates, as in Figure 112.
2 At the centre of each strip draw the mid-ordinate.
3 Determine the lengths of each mid-ordinate.
4 Calculate the average using the relationship:

$$\text{average value} = \frac{\text{sum of the values of the mid-ordinates}}{\text{number of mid-ordinates}}$$

The result given by this method can only be an approximation, but the accuracy is improved by taking a large number of mid-ordinates.

The term *mean value* is sometimes used instead of average value, but they mean the same thing.

For a sinusoidal waveform, we can use a simple formula to obtain the average value instead of using the mid-ordinate rule. This formula can be obtained using calculus or by applying the mid-ordinate rule. We will derive the formula, using the mid-ordinate rule, for a sinusoidal waveform of peak value 1. If we divide the half cycle into six strips, we will have strips of width 30°, as 180° divided by 6 is 30°. The mid-ordinates will thus occur at angles of 15°, 45°, 75°, 105°, 135° and 165°. The value of the sine of these angles can be obtained using tables or a calculator.

θ	$\sin \theta$
15°	0.256
45°	0.707
75°	0.966
105°	0.966
135°	0.707
165°	0.256

The sum of these mid-ordinate values is 3.8636. Hence the average value is

$$\text{average value} = \frac{3.864}{6} = 0.644$$

Using more strips or calculus a more accurate value of 0.637 can be obtained.

If the peak value of the waveform is not 1, the average for the sinusoidal waveform can be obtained by multiplying the peak value by 0.637.

$$\boxed{\text{average} = 0.637 \times \text{peak value}}$$

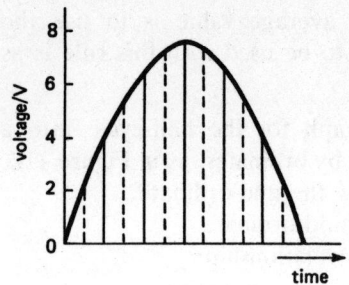

Figure 113

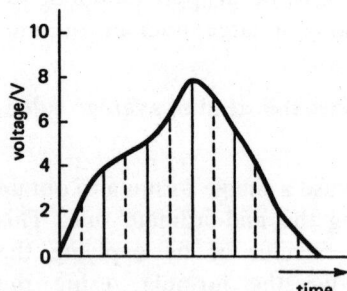

Figure 114

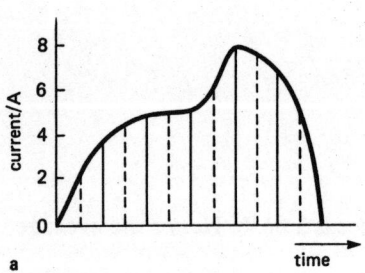

a

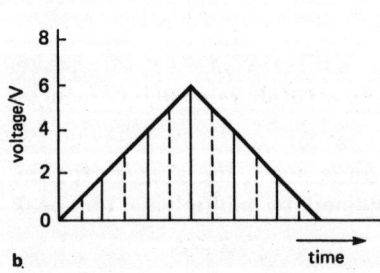

b

Figure 115

Example 7

Determine, using the mid-ordinate rule, the average value of the sinusoidal waveform given in Figure 113.

The waveform has been divided into six strips. The values of the mid-ordinates are: 2 V, 5.2 V, 7.2 V, 7.2 V, 5.2 V, 2 V. Note that the second three strips give, in reverse order, the same values as the first three strips. This is because the waveform is symmetrical.

The sum of the values of the mid-ordinates is 2 + 5.2 + 7.2 + 7.2 + 5.2 + 2 = 28.8 V. Since there are six strips the average value is:

$$\text{average} = \frac{\text{sum of the values of mid-ordinates}}{\text{number of the mid-ordinates}}$$

$$= \frac{28.8}{6} \qquad \text{V}$$

$$= 4.8 \text{ V}$$

Example 8

Determine, using the mid-ordinate rule, the average value of the waveform given in Figure 114.

The waveform has been divided into six strips. The values of the mid-ordinates are: 2.4 V, 4.6 V, 6.2 V, 6.8 V, 4 V, 1 V. The sum of the values of the mid-ordinates is 2.4 + 4.6 + 6.2 + 6.8 + 4 + 1 = 25 V. Since there are six strips the average value is:

$$\text{average} = \frac{\text{sum of the values of the mid-ordinates}}{\text{number of mid-ordinates}}$$

$$= \frac{25.0}{6} \qquad \text{V}$$

$$= 4.2 \text{ V}$$

Example 9

Calculate the average value of a sinusoidal voltage which has a peak value of 5 V.

$$\begin{aligned}
\text{average} &= 0.637 \times \text{peak value} \\
&= 0.637 \times 5 \qquad \text{V} \\
&= 3.185 \text{ V}
\end{aligned}$$

Self-assessment questions

9 Determine, using the mid-ordinate rule, the average values for the waveforms given in Figure 115.

10 Calculate the average value of a sinusoidal current which has a peak value of 2 A.

Root mean square values

When a current i flows through a resistor R the power dissipated is i^2R. This applies even when the current is not direct but is changing. However if the current is changing, the power varies, i.e. the i^2R term varies. The average value of the power dissipated will be the average value of all the i^2R terms. Hence if we take the waveform and divide it into six strips with mid-ordinate values of i_1, i_2, i_3, i_4, i_5 and i_6, then the average power dissipated will be:

$$\text{average} = \frac{\text{sum of } i^2R \text{ mid-ordinate values}}{\text{number of mid-ordinates}}$$

$$= \frac{i_1^2R + i_2^2R + i_3^2R + i_4^2R + i_5^2R + i_5^2R}{6}$$

$$= \frac{(i_1^2 + i_2^2 + i_3^2 + i_4^2 + i_5^2 + i_6^2)R}{6}$$

We could obtain the same power dissipation with a current I through the same resistance if:

$$I^2 = \frac{i_1^2 + i_2^2 + i_3^2 + i_4^2 + i_5^2 + i_6^2}{6}$$

$$I = \sqrt{\left(\frac{i_1^2 + i_2^2 + i_3^2 + i_4^2 + i_5^2 + i_6^2}{6}\right)}$$

But the term in the brackets is the mean of the sum of the squares of the current values. Thus the effective current I is the square root of the mean of the sum of the squares of the current values. This effective current I is known as the root mean square (r.m.s.) current.

A similar discussion can take place with regard to voltages as power dissipated is V^2/R. In general:

> root mean square value = square root of the mean of the sum of the squares of the waveform values

The root mean square value of any waveform can be obtained using the mid-ordinate rule. This applies whether the waveform is regular or irregular.

For a sinusoidal waveform we can derive a root mean square value in a similar way to that used in the previous section for the average value. Thus if we take a sinusoidal waveform of peak value 1 and divide it into six strips, each of angular width 30°, then the mid-ordinate values will be as before and the square of these as in Table 2.

Table 2

θ	sin θ	square of sin θ
15°	0.259	0.067
45°	0.707	0.500
75°	0.966	0.933
105°	0.966	0.933
135°	0.707	0.500
165°	0.259	0.067

Hence the sum of the squares of the waveform values is 3. The root mean square value is therefore:

$$\text{root mean square value} = \sqrt{\left(\frac{3}{6}\right)}$$
$$= 0.707$$

If the peak value of the waveform is not 1 the root mean square value can be obtained, for the sinusoidal waveform, by multiplying the peak value by 0.707.

$$\boxed{\text{root mean square value} = 0.707 \times \text{peak value}}$$

For the average value of a waveform only a half cycle is taken into account, whereas with a root mean square value the entire cycle can be considered. This is because when the negative values for the negative half cycle are squared, the result is a positive quantity. Table 3 gives the data for the negative half cycle of the sinusoidal waveform considered above.

Table 3

θ	sin θ	square of sin θ
195°	−0.259	+0.067
225°	−0.707	+0.500
255°	−0.966	+0.933
285°	−0.966	+0.933
315°	−0.707	+0.500
345°	−0.259	+0.067

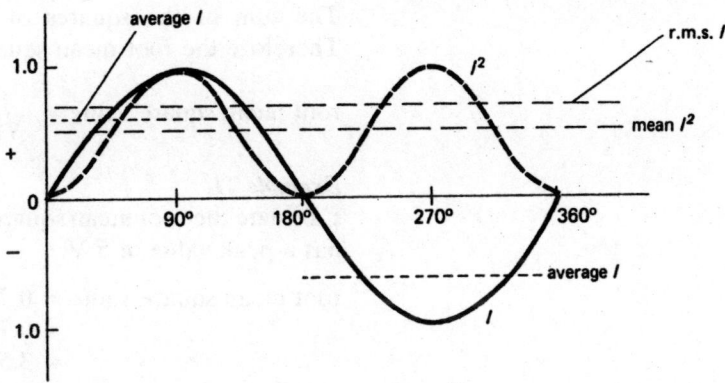

Figure 116

Figure 116 shows how the instantaneous value of a sinusoidal waveform varies with time (angle) and how the squares of these values vary with time. The root mean square value and the average value for the waveform are marked on the graph. Note that the root mean square value and the average value are not the same and that the average can only refer to a half cycle.

Root mean square values of current and voltage may be represented as I_{rms} and V_{rms}, however it is quite common to just represent them as I and V (remember instantaneous values are i and v).

Example 10
Determine, using the mid-ordinate rule, the root mean square value for the waveform given in Figure 114.

This figure was the one used for the calculation of the average value in Example 8. With the waveform divided into six strips the values of the mid-ordinates and their squares are given in Table 4.

Table 4

Mid-ordinate value	Square of mid-ordinate value
2.4	5.76
4.6	21.16
6.2	38.44
6.8	46.24
4.0	16.00
1.0	1.00

The sum of the squares of the mid-ordinate values is 128.6. Therefore the root mean square value is:

$$\text{root mean square value} = \sqrt{\left(\frac{128.6}{6}\right)} = 4.63 \text{ V}$$

Example 11

Calculate the root mean square value of a sinusoidal voltage which has a peak value of 5 V.

$$
\begin{aligned}
\text{root mean square value} &= 0.707 \times \text{peak value} \\
&= 0.707 \times 5.0 \qquad\qquad \text{V}\\
&= 3.54 \text{ V}
\end{aligned}
$$

Self-assessment questions

11 Determine, using the mid-ordinate rule, the root mean square values for the waveforms given in Figure 115.

12 Calculate the root mean square value of a sinusoidal current which has a peak value of 2 A.

Form factor

The form factor is used to give an indication of the shape of a waveform. It is defined as:

$$\text{form factor} = \frac{\text{r.m.s. value}}{\text{average value}}$$

In general the more pointed the peak of the wave concerned the greater the form factor. Thus a sinusoidal waveform has a form factor of 1.11, whereas a square wave has a form factor of 1.

Example 12

What is (a) the r.m.s. value (b) the average value and (c) the form factor for a sinusoidal waveform?

(a) Root mean square value = 0.707 × peak value

(b) Average value = 0.637 × peak value

(c) Form factor $= \dfrac{\text{r.m.s. value}}{\text{average value}}$

$$= \frac{0.707 \times \text{peak value}}{0.637 \times \text{peak value}}$$

$$= 1.11$$

Example 13

What is (a) the r.m.s. value (b) the average value and (c) the form factor for the wave shown in Figure 115(a)?

This is the wave used in self-assessment questions 9 and 11 on page 128, where the mid-ordinate rule had to be used.

(a)   5.33 A

(b)   5.1 A

(c)   Form factor $= \dfrac{\text{r.m.s. value}}{\text{average value}}$

$$= \frac{5.33}{5.1}$$

$$= 1.05$$

Self-assessment questions

13   Calculate the (a) r.m.s. value (b) the average value and (c) the form factor for a square wave.

14   Calculate the (a) r.m.s. value (b) the average value and (c) the form factor, for a triangular wave (like that in Figure 115(b)).

Phasors

Earlier in this chapter we considered the generation of a sinusoidal waveform by the rotation of a line with constant angular velocity, which gives the result shown in Figure 110. This rotating line is known as a phasor.

Figure 117 shows how the phasors and their resulting sinusoidal waveforms appear for two waveforms of the same frequency,

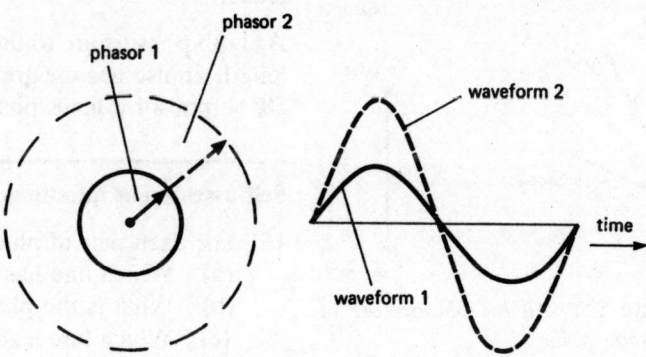

Figure 117 *Two waveforms differing only in peak values*

Figure 118 *Two waveforms differing only by a phase difference*

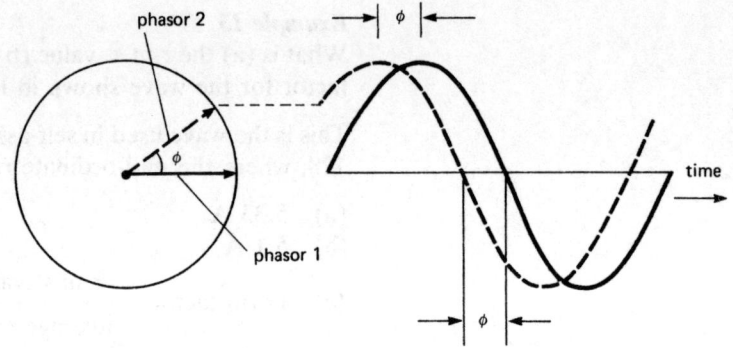

starting at zero values at the same time and differing only in peak
values. The two phasors are always at the same angle regardless of
where on the circular path they are positioned. They differ only in
their length.

Figure 118 shows how the phasors and their resulting sinusoidal
waveforms appear for two waveforms of the same frequency and
the same peak values, but starting with zero values at different
times. The two waveforms are identical in size and form, but one is
displaced from the other by an angle ϕ. This is the angle between
the rotating phasors. This angle is known as the phase angle. In
Figure 118 phasor 2 (and hence waveform 2) leads phasor 1 (and
hence waveform 1) by ϕ. The phasor that first passes through the
horizontal is said to be the leading phasor. Alternatively we could
say that phasor 1 lags phasor 2 by ϕ, since phasor 1 reaches the
horizontal later than phasor 2.

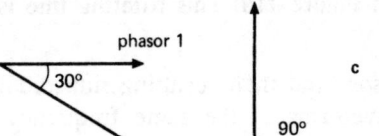

Figure 119 *Both phasors are to the
same scale*

If two waveforms differ in frequency, then the phasors rotate
with different angular velocities (remember that angular velocity
$\omega = 2\pi f$, and so is proportional to the frequency f).

Example 14
For the two phasors shown in Figure 119, which one has the
greatest peak value, what is the phase difference and which one
leads?

As both phasors are to the same scale and phasor 2 has the greater
length, it also has the greatest peak value. The phase difference is
50° so phasor 2 leads phasor 1 by 50°.

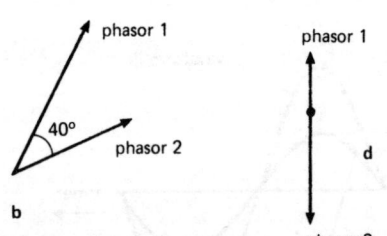

Figure 120 *All the phasors are to
the same scale*

Self-assessment questions

15 For each pair of phasors shown in Figure 120:
   (a) Which one has the greater peak value
   (b) What is the phase difference
   (c) Which one leads?

16 Draw to scale a pair of phasors to represent waveforms which have peak voltages of 2 V and 4 V, when the 4 V waveform leads the other by a phase angle of 30°.

Addition of phasors

In alternating current circuits we may need to find the sum of the potential differences across two components that are in series with each other. The alternating voltage across one component may have a peak value different from the other voltage, and possibly even a phase difference. We cannot obtain the sum by adding the two peak values together or by adding the root mean square values, because the two are not necessarily in phase. We can only add the two peak voltages together to obtain the peak value for the combined waveform when the two are in phase. When they are not in phase the peak voltages of the two waveforms occur at different times. We can, however, obtain the sum of the two by treating the phasors like vectors, taking into account both the size of the phasors and the phase angle between them. In the case of vectors, for example, forces, the sum can be obtained by the parallelogram law. Figure 121 illustrates this approach for phasors.

Phasor 1 gives a waveform that starts at zero time with zero value. Phasor 2 has a greater peak value than phasor 1 and leads it by 90°. The adding of these two phasors gives the result shown in Figure 121. The addition can be drawn graphically, as in the figure. The procedure used is to draw each phasor from the same point, with lengths proportional to the peak values and with the angle between them of the phase angle. The resultant phasor is then obtained by drawing the parallelogram with the two phasors as adjacent sides, and drawing the diagonal through the junction of the two phasors. This diagonal is the resultant phasor and its angles with the other phasors are the phase differences between the resultant phasor and each of the original phasors. In Figure 121 the phase difference between the resultant phasor and phasor 1 is φ.

The same addition can be done algebraically. For the example shown where the two phasors are at right angles, the resultant R can be calculated using Pythagoras's theorem:

$$R^2 = P_1^2 + P_2^2$$

where P_1 and P_2 are the lengths of the two phasors. The phase angle φ is given by:

$$\tan \phi = \frac{P_2}{P_1}$$

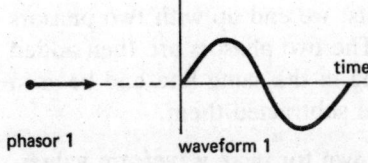

phasor 1 waveform 1

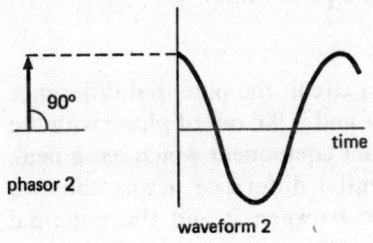

phasor 2 waveform 2

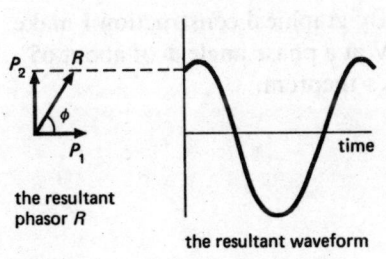

the resultant phasor R the resultant waveform

Figure 121 *Adding two phasors*

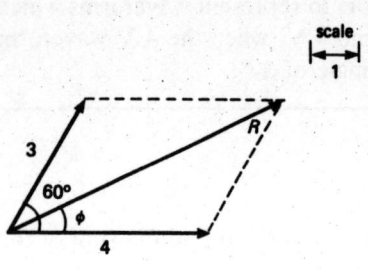

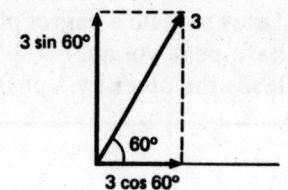

(i) Resolve the 3 phasor into a vertical component of 30 sin 60° and a horizontal component of 3 cos 60°

b

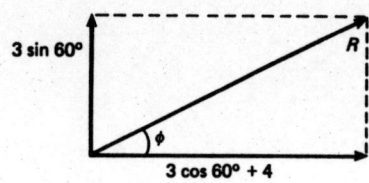

(ii) Combine the horizontal components, and the vertical components. Then use Pythagoras's Theorem to obtain *R*.

Hence $R^2 = (3 \sin 60°)^2 + (3 \cos 60° + 4)^2$ and $\tan \phi = \dfrac{3 \sin 60°}{(3 \cos 60° + 4)}$

a

Figure 122   *Adding phasors*
(a) *Graphical solution* R = 6 *and* φ = 25° *by drawing the parallelogram*
(b) *Algebraic solution* R = 6.08 *and* φ = 25.3°

Where the two phasors are not at right angles, drawing or the algebraic method can be used. In the case of the algebraic method the phasors must be resolved into their horizontal and vertical components before they can be combined using Pythagoras's theorem. Figure 122 shows the graphical method for two such phasors and the way by which the algebraic method is carried out.

In the example shown in Figure 122, after resolving a phasor into horizontal and vertical components, we end up with two phasors both in the horizontal direction. The two phasors are then added together. If the two phasors acting in the same line had been in opposite directions we would have subtracted them.

As seen above, phasors can be drawn for peak waveform values. They can also be used for r.m.s. values. This is because the r.m.s. value is directly proportional to the peak value.

Example 15
For two components in series in a circuit the potential difference across one has a peak value of 2 V and is 90° out of phase with the potential difference across the other component which has a peak value of 4 V. What is the potential difference across the two components and the phase angle between it and the potential difference across the first component?

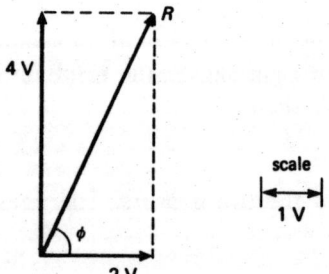

Figure 123

Figure 123 shows this situation. By graphical construction I make the resultant phasor *R* to be 4.5 V at a phase angle φ of about 65°. By calculation, using Pythagoras's theorem:

$$R^2 = 4^2 + 2^2$$
$$R = 4.47 \text{ V}$$
$$\tan \phi = \frac{4}{2}$$
$$\phi = 63.4°$$

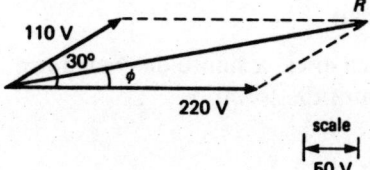

Figure 124

Example 16

Two sources of alternating e.m.f. are connected in series. If one has an e.m.f. of 220 V and the other 110 V and there is a phase angle of 30° between them, what is the total e.m.f.?

Figure 124 shows this situation. By graphical construction I make the resultant phasor R, the total e.m.f., to be 320 V at a phase angle to the 220 V of about 10°. By calculation we have:

Vertical component of 110 V is 110sin 30°.
Horizontal component of 110 V is 110cos 30°.
Hence total vertical components = 110sin 30°.
Total horizontal components = 110cos 30° + 220

Hence, using Pythagoras's theorem:
$$R^2 = (110\sin 30°)^2 + (110\cos 30° + 220)^2$$
$$R = 320 \text{ V}$$
$$\tan \phi = \frac{110\sin 30°}{110\cos 30° + 220}$$
$$\phi = 9.9°$$

Self-assessment questions

17 For two components in series the potential difference across one has a peak value of 5 V and that across the other, 3 V, and there is a phase difference of 45° between the two. What is the total potential difference across the two components?

18 For two components in series the potential difference across one component has a root mean square value of 20 V and that across the other, 10 V and there is a phase difference of 90° between the two. What is the total root mean square potential difference across the two components?

19 Two sources of alternating e.m.f. are connected in series. One has a peak e.m.f. of 12 V and the other 6 V and there is a phase difference of 60° between the two. What is the total e.m.f.?

20 Two sinusoidal currents flow through the same component. One has a peak value of 2 A and the other 5 A. If the 2 A current leads the other by a phase angle of 70°, what is the peak value of the total current and its phase angle with respect to the 5 A current? Both currents have the same frequency.

The sinusoidal waveform equation

As has been shown earlier in this chapter, a sinusoidal waveform can be represented by an equation of the form:

$$v = V_m \sin \theta$$

or: $v = V_m \sin \omega t$

where v is the instantaneous value of the waveform at time t, V_m is the peak value and ω the angular velocity, i.e. the rate at which angle θ varies with time t.

The above equation gives a graph, as in Figure 110, which starts with a zero value of v at time $t = 0$. We can, however, have sinusoidal waveforms which are at some angle ϕ at time $t = 0$. To take account of this the above equations have to be modified to:

$$v = V_m \sin (\theta + \phi)$$
$$v = V_m \sin (\omega t + \phi)$$

Example 17

Write down the equation for the sinusoidal voltage which has a peak value of 4 V and a frequency of 50 Hz if at zero time the voltage is zero.

As $\omega = 2\pi f$
$$\omega = 2\pi \times 50 = 100\pi$$
Hence: $v = V_m \sin \omega t$
$$= 4 \sin 100\pi\, t$$

Note there is no angle ϕ because the voltage is zero at time $t = 0$.

Example 18

Write down the equation for the sinusoidal voltage which has a peak value of 5 V and a frequency of 50 Hz if it leads the voltage in the previous example by $\pi/2$.

$\omega = 2\pi f = 2\pi \times 50 = 100\pi$, $\phi = \pi/2$.
Hence: $v = V_m \sin (\omega t + \phi)$
$$= 5 \sin (100\pi\, t + \pi/2)$$

Example 19

Write down the equation for the sinusoidal voltage which has a peak value of 1 V and frequency of 50 Hz if it lags the voltage in Example 17 by $\pi/2$.

$\omega = 2\pi f = 2\pi \times 50 = 100\pi$, $\phi = -\pi/2$
Hence: $v = V_m \sin (\omega t + \phi)$
$$= 1 \sin (100\pi\, t - \pi/2)$$

Self-assessment question

21 Write down the equations for the following sinusoidal waveforms if they all have a frequency of 50 Hz.

(a) A voltage of peak value 2 V which has zero voltage at zero time.

(b) A voltage of peak value 4 V which leads the voltage in (a) by 90°.

(c) A voltage of peak value 3 V which lags the voltage in (a) by $\pi/4$.

(d) A current of peak value 1.5 A and in phase with the voltage in (a).

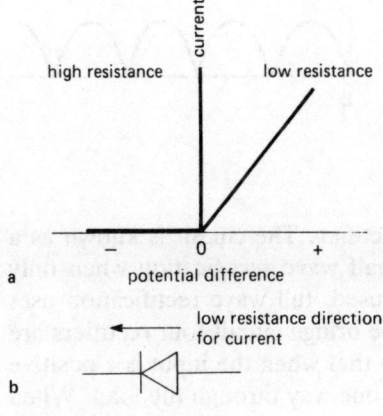

Figure 125 *A rectifier*
(a) *Characteristic*
(b) *Symbol*

Rectification

Rectification is the conversion of an alternating voltage to a unidirectional voltage. Rectification can be achieved by the use of a component that has a low resistance for current flow through it in one direction but a very high resistance for current flow in the reverse direction. Such devices are diode valves, semiconductor diodes and metal rectifiers. Figure 125 shows the basic characteristic, i.e. voltage-current relationship, for a device of this type.

If a rectifier is inserted in series with the load in a circuit, as in Figure 126, the current through the load becomes unidirectional although the input is an alternating voltage. Such a circuit is said to be a half wave rectifier. This is because only the positive half cycles of the input find a low resistance path. It is as though the rectifier switches on and allows the current through for the positive half cycles but switches off for the negative half cycles.

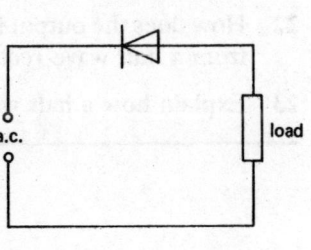
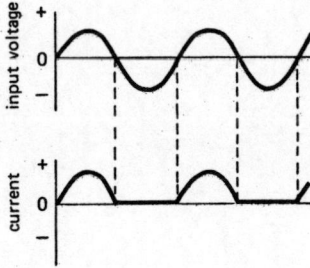

Figure 126 *Half wave rectification*

Figure 127   *Full wave rectification*
(a) *The circuit*
(b) *The path for a positive half cycle*
(c) *The path for a negative half cycle*
(d) *The current through the load*

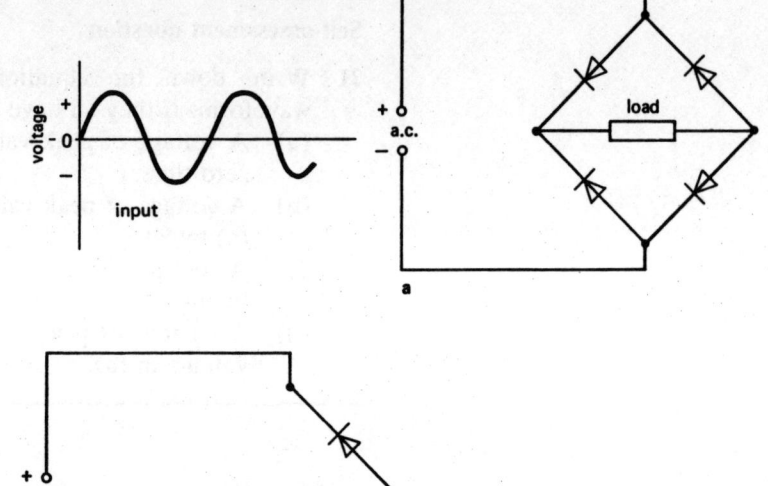

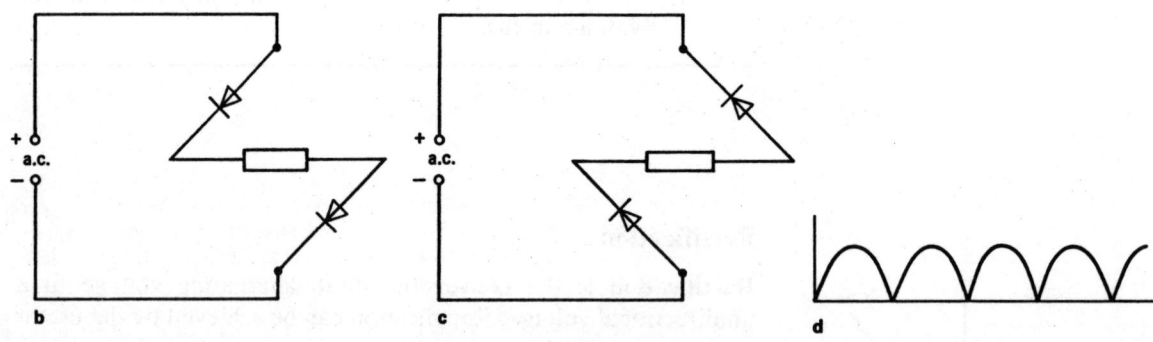

Figure 127 shows a full wave rectifier. The circuit is known as a bridge rectifier circuit. Unlike half wave rectification when only the positive half of the cycle is used, full wave rectification uses both halves of the cycle. With the bridge circuit four rectifiers are used and arranged in such a way that when the input is a positive half cycle the current is directed one way through the load. When the input is the negative half cycle a different current path is used so that the current flow through the resistor remains in the same direction.

Self-assessment questions

22   How does the output from a full wave rectifier differ from that from a half wave rectifier?

23   Explain how a half wave rectifier circuit operates.

---

7 Single phase a.c. circuits

After reading this chapter you should be able to:

1 Solve a.c. circuit problems where there is a purely resistive circuit, a purely capacitive circuit or a purely inductive circuit.
2 Explain the phase relationships for the circuits.
3 Explain the term reactance and calculate inductive and capacitive reactances.
4 Solve a.c. circuit problems involving series combinations of:
   (a) inductance and resistance
   (b) capacitance and resistance
   (c) inductance, capacitance and resistance.
5 Explain the term impedance and calculate impedances for series circuits.
6 Describe and use voltage and impedance triangles.
7 Show that the average power in an a.c. circuit depends only on the resistance.
8 Explain the terms true power, apparent power, reactive power and power factor, solving problems involving them.
9 Derive from a phasor diagram the resonance condition for a series LCR circuit and explain the conditions that occur.
10 Explain what is meant by voltage magnification.

A pure resistor in an a.c. circuit

Figure 128 shows an a.c. circuit where there is only pure resistance. The term 'pure' is used to signify that there are no inductive or capacitive effects. With a real resistor there would be invariably some self inductance.

With a pure resistance the current through the resistance is in phase with the potential difference across it.

Figure 129 shows the phasors and the sinusoidal waveforms for the current and potential difference.

We can show that the current and potential difference are in phase by considering the current to be:

$$i = I_m \sin \omega t$$

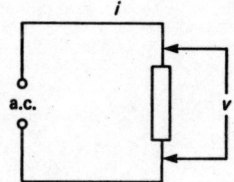

Figure 128

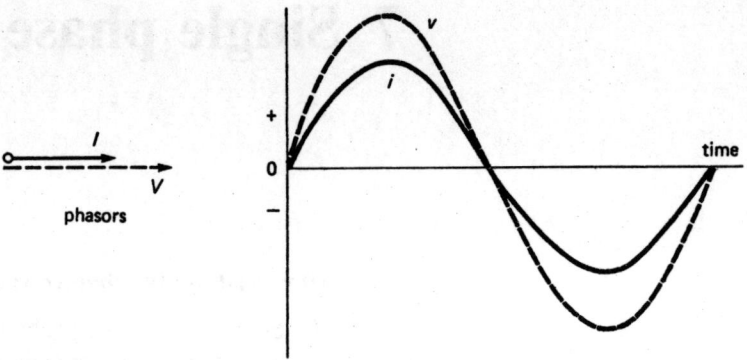

Figure 129   *Circuit containing pure resistance*

As for the instantaneous value of v and i, we have $v = iR$, so:

$v = Ri = RI_m\sin \omega t$

But RI_m is the peak value of the potential difference V_m, hence:

$v = V_m\sin \omega t$

There is no phase difference term in this equation compared with the current equation, so they are in phase.

For the instantaneous currents and potential difference we have $R = v/i$. Also for the peak values, because they occur at the same time, we have $R = V_m/I_m$. As the root mean square current $I_{rms} = I_m/\sqrt{2}$ and $V_{rms} = V_m/\sqrt{2}$, then:

$$R = \frac{V_m}{I_m} = \frac{\sqrt{2} \times I_{rms}}{\sqrt{2} \times V_{rms}}$$

$$R = \frac{I_{rms}}{V_{rms}}$$

Thus $R = \dfrac{v}{i} = \dfrac{V_m}{I_m} = \dfrac{V_{rms}}{I_{rms}}$

See Chapter 6 for the way in which root mean square current is defined. This shows that the power dissipated by an alternating current through a pure resistor is given by:

$$\boxed{\text{power} = I_{rms}^2 R}$$

Example 1

An alternative current of peak value 2 A passes through a pure resistor of 100 Ω. What is (a) the peak potential difference across the resistor and (b) the power dissipated in that resistor?

(a) $I_m = 2$ A, $R = 100$ Ω

$$R = \frac{V_m}{I_m}$$

Hence: $V_m = R \, I_m$

$\qquad = 100 \times 2$ $\qquad\qquad\qquad\qquad$ $\Omega \times$ A

$\qquad = 200$ V

(b) $\qquad\qquad I_{rms} = \frac{I_m}{\sqrt{2}}$

$\qquad\qquad\qquad = \frac{2}{\sqrt{2}}$ $\qquad\qquad\qquad\qquad$ A

$\qquad\qquad\qquad = 1.41$ A

$\qquad\qquad$ power $= I_{rms}^2 R$

$\qquad\qquad\qquad = 1.41^2 \times 100$ $\qquad\qquad\qquad$ $A^2 \times \Omega$

$\qquad\qquad\qquad = 200$ W

Example 2

An alternating voltage of root mean square value 4 V is applied across a pure resistor of 10 Ω. What is:

(a) The peak potential difference across the resistor
(b) The peak current through the resistor
(c) The root mean square current
(d) The power dissipated in the resistor?

(a) $V_{rms} = 4$ V, $R = 10$ Ω

$\qquad\qquad V_m = V_{rms} \times \sqrt{2}$

$\qquad\qquad\qquad = 4 \times \sqrt{2}$ $\qquad\qquad\qquad\qquad$ V

$\qquad\qquad\qquad = 5.66$ V

(b) $\qquad\qquad R = \frac{V_m}{I_m}$

Hence $I_m = \frac{V_m}{R}$

$\qquad\qquad\qquad = \frac{5.66}{10}$ $\qquad\qquad\qquad\qquad\qquad$ $\dfrac{V}{\Omega}$

$\qquad\qquad\qquad = 0.566$ A

(c) $\qquad I_{rms} = \dfrac{I_m}{\sqrt{2}}$

$\qquad\qquad = \dfrac{0.566}{\sqrt{2}}$ A

$\qquad\qquad = 0.4\ A$

We could have obtained this value by using the equation:

$$R = \dfrac{V_{rms}}{I_{rms}}$$

$\qquad I_{rms} = \dfrac{4}{10}$ $\dfrac{V}{\Omega}$

$\qquad\qquad = 0.4\ A$

(d) $\qquad$ power $= I_{rms}^2 R$

$\qquad\qquad = 0.4^2 \times 10$ $A^2 \times \Omega$

$\qquad\qquad = 1.6\ W$

We could have obtained this value by using the equation:

$$power = \dfrac{V_{rms}^2}{R}$$

$\qquad\qquad = \dfrac{4^2}{10}$ $\dfrac{V^2}{\Omega}$

$\qquad\qquad = 1.6\ W$

Self-assessment questions

1   What is the phase relationship between the alternating current through a pure resistor and the alternating potential difference across it?

2   An alternating current of peak value 100 mA passes through a pure resistor of 20 Ω. What is:
   (a) The peak potential difference across the resistor
   (b) The root mean square potential difference
   (c) The power dissipated in that resistor?

3   An alternating voltage of peak value 24 V is applied across a pure resistor of 40 Ω. What is:
   (a) The root mean square potential difference across the resistor
   (b) The root mean square current through it
   (c) The power dissipated in the resistor?

4 An alternating voltage of root mean square value 240 V is applied across a pure resistor. If the power dissipated in that resistor is 1 kW, what is:

(a) The value of the resistance

(b) The root mean square current

(c) The peak potential difference

(d) The peak current?

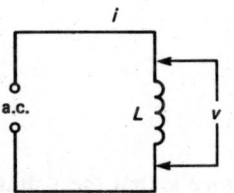

Figure 130

A pure inductor in an a.c. circuit

A pure inductor is a component that has only inductance and no resistance or capacitance. As an inductor is essentially just a coil of wire, perhaps wound on a ferromagnetic former, it will have resistance. The effect of the resistance on the performance of an inductor in a circuit is considered later in this chapter.

Figure 130 shows the basic circuit with an alternating potential difference applied across the inductor. If the current is considered to be represented by the equation:

$$i = I_m \sin \omega t$$

then the induced e.m.f. in the component due to self inductance is:

$$E = L \times \text{rate of change of current}$$

where L is the self inductance. This induced e.m.f. is in the opposite direction to the applied e.m.f. Hence, applying Kirchhoff's laws:

$$\text{applied e.m.f.} - \text{induced e.m.f.} - iR = 0$$

But R is zero, hence:

$$\text{applied e.m.f.} = \text{induced e.m.f.}$$

Since $v = E$, the applied e.m.f., i.e. the potential difference v across the inductor in this case, is given by:

$$v = L \times \text{rate of change of current}$$

Figure 131 shows how the alternating current varies with time and how the potential difference v must be a maximum when the rate of change of current with time is a maximum and zero when the rate of change of current with time is zero. As will be apparent from the resulting potential difference–time graph, the potential difference is out of phase with the current.

The current lags the voltage by 90° (or π/2).

The rate of change of current with time can be obtained by using calculus and differentiating the current equation. The result is:

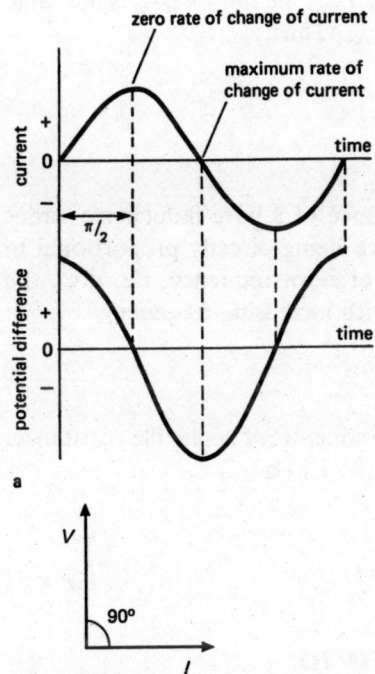

Figure 131 *Circuit containing pure inductance*
(a) *The waveforms*
(b) *The phasors*

rate of change of current $= \omega I_m \cos \omega t$

Hence: $\quad v = \omega L I_m \cos \omega t$

$\qquad\qquad v = V_m \cos \omega t$

where $V_m \;\; = \omega L I_m$

Thus: $\quad \dfrac{V_m}{I_m} = \omega L$

The quantity V/I in a d.c. circuit is called the resistance, in an inductive circuit the quantity V_m/I_m is called the reactance X_L. Thus:

$$X_L = \omega L = 2\pi\, fL$$

The unit of reactance is ohms.

An important point to note with reactance is that the voltage that is being divided by the current does not occur at the same time as the current value concerned. It is not the instantaneous voltage divided by the current at that instant.

Since $V_m = \sqrt{2}\, V_{\text{rms}}$ and $I_m = \sqrt{2}\, I_{\text{rms}}$, the ratio V_m/I_m is the same as $V_{\text{rms}}/I_{\text{rms}}$. Hence $X_L = V_{\text{rms}}/I_{\text{rms}}$. Thus:

$$X_L = \frac{V_m}{I_m} = \frac{V_{\text{rms}}}{I_{\text{rms}}}$$

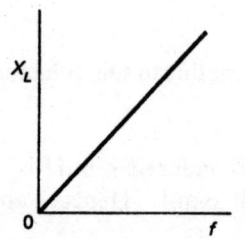

Figure 132 *Variation of reaction with frequency for pure inductance*

Figure 132 shows how the reactance of a pure inductance varies with frequency with the reactance being directly proportional to the frequency. This means that at zero frequency, i.e. d.c., the reactance is zero but increases with increasing frequency.

Example 3

What is the reactance of a 0.2 H inductor, of negligible resistance, at a frequency of (a) 50 Hz and (b) 1 kHz?

(a) $\quad L = 0.2$ H, $f = 50$ Hz

$\qquad X_L = 2\pi\, fL$

$\qquad\qquad = 2\pi \times 50 \times 0.2 \qquad\qquad$ Hz × H

$\qquad\qquad = 31.4\ \Omega$

(b) $\quad L = 0.2$ H, $f = 1$ kHz $= 1000$ Hz

$\qquad X_L = 2\pi\, fL \qquad\qquad\qquad\qquad\qquad$ Hz × H

$\qquad\qquad = 2\pi \times 1000 \times 0.2$

$\qquad\qquad = 1257\ \Omega$

Note that the higher the frequency the greater the reactance of an inductor.

Example 4

A sinusoidal voltage of root mean square value 6 V and frequency 50 Hz is applied to an inductor of negligible resistance. If the root mean square current is 12 mA, what is the inductance of the inductor?

$V_{rms} = 6$ V, $f = 50$ Hz, $I_{rms} = 12$ mA $= 12 \times 10^{-3}$ A

$$X_L = \frac{V_{rms}}{I_{rms}}$$

$$= \frac{6}{12 \times 10^{-3}} \qquad \frac{V}{A}$$

$$= 500 \ \Omega$$

$$X_L = 2\pi fL$$

Hence: $L = \dfrac{X_L}{2\pi f}$

$$= \frac{500}{2\pi \times 50} \qquad \frac{\Omega}{Hz}$$

$$= 1.59 \text{ H}$$

Self-assessment questions

5   What is the phase relationship between the alternating current through a pure inductor and the alternating potential difference across it?

6   Define reactance for an inductor.

7   How does the reactance of an inductor change with frequency?

8   What is the reactance of a 50 mH inductor, of negligible resistance, at a frequency of (a) 50 Hz and (b) 50 kHz?

9   A sinusoidal voltage of root mean square value 12 V and frequency 1 kHz is applied to an inductor of negligible resistance. If the root mean square current is 20 mA, what is the inductance of the inductor?

10   What is the reactance of an inductor if a root mean square potential difference of 4 V across it gives rise to a root mean square current of 12 mA through it?

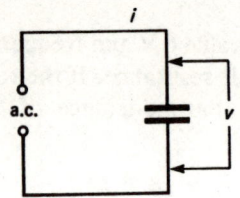

Figure 133

A pure capacitor in an a.c. circuit

A pure capacitor is a device which has capacitance but no resistance or inductance. Figure 133 shows the basic circuit in which an alternating potential difference is applied across a capacitor. For a sinusoidal potential difference we have:

$$v = V_m \sin \omega t$$

The charge q on the capacitor is related to the potential difference v by the equation:

$$q = Cv$$

where C is the capacitance. Thus:

$$q = CV_m \sin \omega t$$

This equation shows how the charge q on a capacitor plate varies with time. But current is the rate of movement of charge, thus:

$$i = \text{rate of change of } q$$
$$i = \text{rate of change of } (CV_m \sin \omega t)$$

By differentiating this term we can arrive at the rate of change, which is:

$$i = \omega CV_m \cos \omega t$$

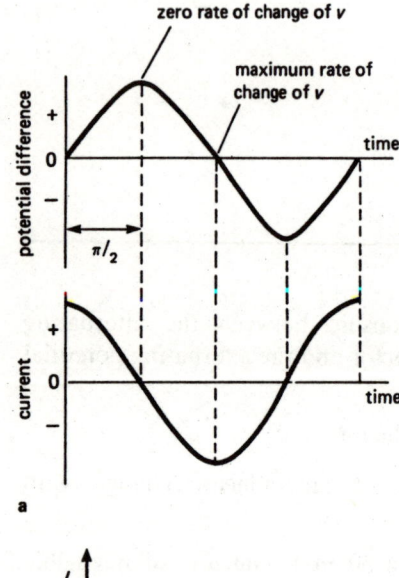

Figure 134 shows the graphs of the potential difference and the current with respect to time.

The current leads the voltage by 90° (or π/2).

This is illustrated by the phasor diagram given in the figure.

The above equation for the current can be written as:

$$i = I_m \cos \omega t$$

where $I_m = \omega CV_m$.

Thus: $$\frac{V_m}{I_m} = \frac{1}{\omega C}$$

The quantity V_m/I_m is called the reactance X_C. Thus:

$$\boxed{X_C = \frac{1}{\omega C} = \frac{1}{2\pi fC}}$$

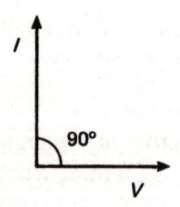

Figure 134 *Circuit containing pure capacitance*
(a) *The waveforms*
(b) *The phasors*

The unit of the reactance is the ohm.

An important point to note is that the voltage that is being divided by the current does not occur at the same time as the current value concerned. It is not the instantaneous voltage divided by the current at that instant.

Since $V_m = \sqrt{2}\,I_{rms}$ and $I_m = \sqrt{2}\,I_{rms}$, the ratio V_m/I_m is the same as V_{rms}/I_{rms}. Hence:

$$X_C = \frac{V_m}{I_m} = \frac{V_{rms}}{I_{rms}}$$

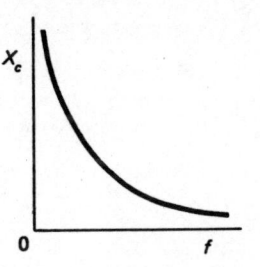

Figure 135 shows how the reactance of a pure capacitor varies with frequency, the reactance being inversely proportional to the frequency. This means that at zero frequency the reactance is infinite for a capacitor and becomes smaller as the frequency increases.

Figure 135 *Variation of reactance with frequency for pure capacitance*

Example 5

What is the reactance of a 8 μF capacitor at a frequency of (a) 50 Hz and (b) 1 kHz?

(a) $C = 8\ \mu F = 8 \times 10^{-6}\ F, f = 50\ Hz$

$$X_C = \frac{1}{2\pi fC}$$

$$= \frac{1}{2\pi \times 50 \times 8 \times 10^{-6}} \qquad \frac{1}{Hz \times F}$$

$$= 398\ \Omega$$

(b) $C = 8\mu F = 8 \times 10^{-6}\ F, f = 1\ kHz = 1000\ Hz$

$$X_C = \frac{1}{2\pi fC}$$

$$= \frac{1}{2\pi \times 1000 \times 8 \times 10^{-6}} \qquad \frac{1}{Hz \times F}$$

$$= 19.9\ \Omega$$

Note that the higher the frequency the smaller the reactance of the capacitor.

Example 6

A sinusoidal voltage of root mean square value 6 V and frequency 50 Hz, is applied to a capacitor. If the root mean square current is 1 mA, what is the capacitance of the capacitor?

$$V_{rms} = 6 \text{ V}, f = 50 \text{ Hz}, I_{rms} = 1 \text{ mA} = 1 \times 10^{-3} \text{ A}$$

$$X_C = \frac{V_{rms}}{I_{rms}}$$

$$= \frac{6}{1 \times 10^{-3}} \qquad \frac{\text{V}}{\text{A}}$$

$$= 6000 \ \Omega$$

$$X_C = \frac{1}{2\pi fC}$$

Hence: $$C = \frac{1}{X_C \times 2\pi f}$$

$$= \frac{1}{6000 \times 2\pi \times 50} \qquad \frac{1}{\Omega \times \text{Hz}}$$

$$= 5.31 \times 10^{-7} \text{ F}$$
$$= 0.531 \ \mu\text{F}$$

Self-assessment questions

11 What is the phase relationship between the alternating current in a circuit containing only capacitance and the alternating potential difference?

12 Define reactance for a capacitor.

13 How does the reactance of a capacitor change with frequency?

14 What is the reactance of a 16 μF capacitor at a frequency of (a) 50 Hz and (b) 50 kHz?

15 A sinusoidal voltage of root mean square value 20 V and frequency 1 kHz is applied to a capacitor. If the root mean square current in the circuit is 4 mA, what is the capacitance of the capacitor?

16 What is the reactance of a capacitor if a root mean square potential difference across it of 10 V gives rise to a root mean square current of 0.5 mA through it?

Inductance and resistance in series

A real inductor, as opposed to the pure inductor considered earlier in this chapter, will have both inductance and resistance. It

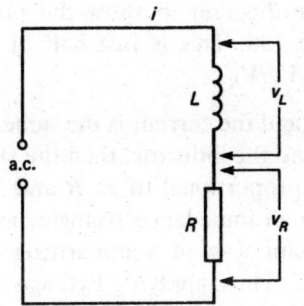

Figure 136 *An inductor in series with a resistor*

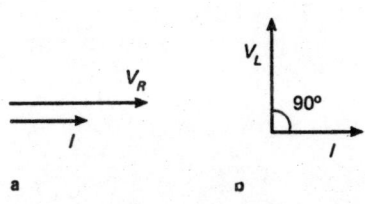

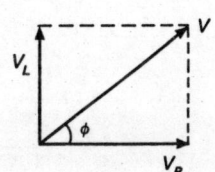

Figure 137 *Phasors for an inductor in series with a resistor*
(a) *Resistor*
(b) *Inductor*
(c) *Adding the voltage phasors*

behaves like a pure inductor in series with a pure resistor. The following discussion applies to a real inductor, as well as to an inductor which is in series with a resistor.

Figure 136 shows the basic circuit. The total potential difference across the two components cannot be obtained by adding together the potential differences across each component because the two are not in phase. The potential difference across the resistor is in phase with the current through it, but the potential difference across the inductor leads the current by 90°. Figure 137 shows the two potential difference phasors. The total potential difference across the two components V is given by:

$$V^2 = V_L^2 + V_R^2$$

The potential differences can be all considered either as peak values or as root mean square values. But $V_L = X_L I$ and $V_R = RI$, hence:

$$V^2 = X_L^2 I^2 + R^2 I^2$$
$$V^2 = I^2(X_L^2 + R^2)$$
$$\frac{V}{I} = \surd(X_L^2 + R^2)$$

V is the total potential difference and I the current through the circuit. The ratio V/I is a measure of the impedance to the current flow of the components concerned (for a d.c. circuit V/I is a measure of the resistance to the current flow and is known as the resistance). The ratio V/I, which is the ratio of peak values or the ratio of root mean square values, is defined as being the circuit impedance Z.

$$\boxed{Z = \frac{V}{I}}$$

The unit of impedance is the ohm. Thus for the inductor in series with the resistance the circuit impedance is given by:

$$\boxed{Z = \surd(X_L^2 + R^2)}$$

As the parallelogram in Figure 137(c) indicates, the total potential difference V leads the potential difference across the resistor, and hence the current, by an angle ϕ, where:

$$\boxed{\tan \phi = \frac{V_L}{V_R}}$$

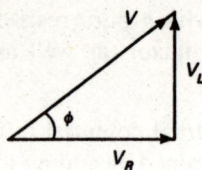

Figure 138   *Voltage triangle*

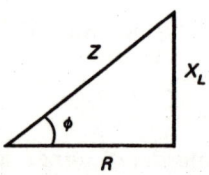

Figure 139   *Impedance triangle*

An alternative way of drawing the diagram to show the phase angle is as a triangle, as in Figure 138. This is just half of the parallelogram. As before, $\tan \phi = V_L/V_R$.

As $V = IZ$, $V_R = IR$ and $V_L = IX_L$ and the current is the same for the circuit as a whole, the resistor and the inductor, then the sides of the voltage triangle are directly proportional to Z, R and X_L. The triangle can thus be drawn as an impedance triangle, as in Figure 139. This triangle is a useful way of summarizing the various relationships for impedances. Thus, applying Pythagoras's theorem to the triangle gives:

$$Z^2 = X_L^2 + R^2$$

Also considering the various trigonometrical relationships gives:

$$\tan \phi = \frac{X_L}{R}$$

$$\sin \phi = \frac{X_L}{Z}$$

$$\cos \phi = \frac{R}{Z}$$

Example 7
An inductor has a resistance of 100 Ω and an inductance of 0.5 H. What is its impedance at a frequency of 1 kHz?

$R = 100\ \Omega$, $L = 0.5$ H, $f = 1$ kHz $= 1000$ Hz
$$X_L = 2\pi fL$$
$$= 2\pi \times 1000 \times 0.5 \qquad\qquad\qquad \text{Hz} \times \text{H}$$
$$= 3142\ \Omega$$
$$Z = \sqrt{(X_L^2 + R^2)}$$
$$= \sqrt{(3142^2 + 100^2)}$$
$$= 3144\ \Omega$$

Example 8
An inductor has an impedance of 200 Ω. What will be the root mean square current through it when the potential difference across it is 6 V root mean square?

$Z = 200\ \Omega$, $V_{rms} = 6$ V
$$Z = \frac{V_{rms}}{I_{rms}}$$

Hence: $I_{rms} = \dfrac{V_{rms}}{Z}$

$= \dfrac{6}{200}$ $\dfrac{V}{\Omega}$

$= 0.03$ A

Example 9

An inductor with a resistance of 50 Ω and an inductance of 0.2 H is connected to a 50 Hz, 240 V root mean square voltage supply. Calculate the root mean square current in the circuit and the phase angle between the potential difference across the inductor and the current.

$R = 50\ \Omega,\ L = 0.2$ H, $f = 50$ Hz, $V_{rms} = 240$ V

$X_L = 2\pi fL$

    $= 2\pi \times 50 \times 0.2$ Hz $\times$ H

    $= 62.8\ \Omega$

$Z = \sqrt{(X_L^2 + R^2)}$

   $= \sqrt{(62.8^2 + 50^2)}$ Ω

   $= 80.3\ \Omega$

The relationship $Z = V/I$ applies where V and I are either both peak values or both root mean square values. Thus:

$$Z = \dfrac{V_{rms}}{I_{rms}}$$

Hence: $I_{rms} = \dfrac{V_{rms}}{Z}$

$= \dfrac{240}{80.3}$ $\dfrac{V}{\Omega}$

$= 2.99$ A

Any one of the trigonometrical relationships obtained from the impedance triangle can be used to obtain ϕ. Thus, for example,

$\tan \phi = \dfrac{X_L}{R}$

$= \dfrac{80.3}{50}$ $\dfrac{\Omega}{\Omega}$

$\phi = 58.1°$

This is the angle by which the potential difference across the 'impure' inductor leads the current through it.

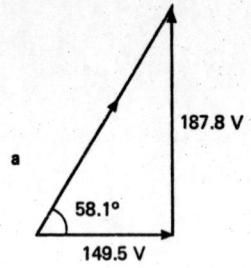

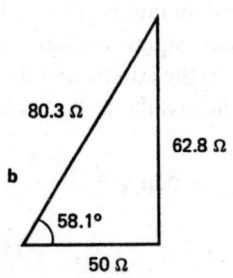

Figure 140   (a) *Voltage triangle*
(b) *Impedance triangle*

Example 10
Draw the voltage and impedance triangles for the circuit described in Example 9.

For the voltage triangle:
$$V_R = IR = 2.99 \times 50 \qquad A \times \Omega$$
$$= 149.5 \text{ V}$$
$$V_L = IX_L = 2.99 \times 62.8 \qquad A \times \Omega$$
$$= 187.8 \text{ V}$$

The resulting voltage triangle is shown in Figure 140(a).

For the impedance triangle:

$$X_L = 62.8 \ \Omega \quad \text{and} \quad R = 50 \ \Omega$$

The resulting impedance triangle is shown in Figure 140(b). Both triangles are exactly the same shape, since all corresponding triangles are the same. If the appropriate scale is chosen corresponding sides are the same lengths.

Example 11
A coil takes a current of 0.4 A when a direct potential difference of 10 V is connected across it. When an alternating potential difference of root mean square value 20 V and frequency 50 Hz is applied the root mean square current is 0.1 A. What is (a) the resistance and (b) the inductance, of the coil?

For the d.c.: $I = 0.4$ A, $V = 10$ V

$$R = \frac{V}{I}$$

$$= \frac{10}{0.4} \qquad \frac{V}{A}$$

$$= 25 \ \Omega$$

For the a.c.: $I_{\text{rms}} = 0.1$ A, $V_{\text{rms}} = 20$ V, $f = 50$ Hz

$$Z = \frac{V_{\text{rms}}}{I_{\text{rms}}}$$

$$= \frac{20}{0.1} \qquad \frac{V}{A}$$

$$= 200 \ \Omega$$
$$Z^2 = X_L^2 + R^2$$
Hence: $$X_L^2 = Z^2 - R^2$$
$$= 200^2 - 25^2$$
$$X_L = 198.4 \ \Omega$$
$$X_L = 2\pi fL$$

Hence: $L = \dfrac{X_L}{2\pi f}$ Ω

$= \dfrac{198.4}{2\pi \times 50}$ Hz

$= 0.632$ H

Self-assessment questions

17 An inductor has a resistance of 50 Ω and an inductance of 0.2 H. What is its impedance at a frequency of 50 Hz?

18 An inductor has an impedance of 1 kΩ. What will be the root mean square current through it when the root mean square potential difference across it is 10 V?

19 An inductor has a resistance of 60 Ω and an inductance of 0.4 H. If it is connected to a 240 V root mean square voltage supply at 50 Hz, what will be the root mean square current? What is the phase angle between the potential difference across the inductor and the current?

20 A coil has a root mean square current of 3.0 A passing through it when a root mean square potential difference of 150 V is connected across it. What is the impedance of the coil? What would happen to the impedance if the frequency of the supply was increased?

21 A coil has a root mean square current of 60 mA passing through it when a root mean square voltage of 20 V is connected across it and the frequency of the voltage is 100 Hz. The same coil has a current of 120 mA passing through it when a direct voltage of 6 V is connected across it. What is (a) the inductance and (b) the resistance, of the coil?

22 Draw a voltage and an impedance triangle for an inductor which has an inductance of 2 H and a resistance of 200 Ω in a circuit where a root mean square potential difference of 20 V is applied across it. The frequency is 100 Hz. State the phase angle.

23 A coil of resistance 20 Ω and inductance 0.3 H is connected in series with a resistor of resistance 30 Ω and a root mean square voltage of 120 V, at a frequency of 50 Hz, connected across the two components. Determine the potential differences across (a) the coil and (b) the resistor.

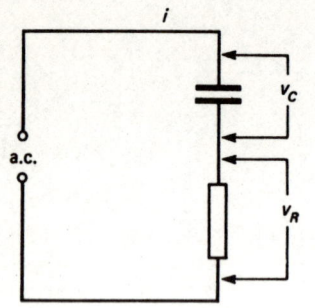

Figure 141   *A capacitor in series with a resistor*

Capacitor and resistor in series

Figure 141 shows the basic circuit. For a resistor the potential difference across it, V_C, is in phase with the current through it. For a capacitor the current leads the potential difference V_C by 90°. The total potential difference across the two components cannot therefore be obtained by adding together V_R and V_C because the two are not in phase. Figure 142 shows the phasors for this situation. The total potential difference across the two components V is given by:

$$V^2 = V_C^2 + V_R^2$$

The potential differences are either all peak values or all root mean square values. But $V_C = X_C I$ and $V_R = RI$, hence:

$$V^2 = X_C^2 I^2 + R^2 I^2$$
$$V^2 = I^2(X_C^2 + R^2)$$
$$\frac{V}{I} = \sqrt{(X_C^2 + R^2)}$$

The ratio V/I is the impedance Z of the circuit. Thus:

$$\boxed{Z = \frac{V}{I} = \sqrt{(X_C^2 + R^2)}}$$

The unit of impedance is the ohm.

As the parallelogram in Figure 142 indicates, the total potential difference V lags the potential difference across the resistor, and hence the current, by an angle ϕ, where:

$$\tan \phi = \frac{X_C}{R}$$

An alternative way of drawing the diagram to show the phase angle is as a triangle, as in Figure 143. This is just half the parallelogram. As before, $\tan \phi = V_C/V_R$.

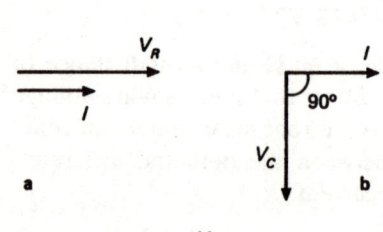

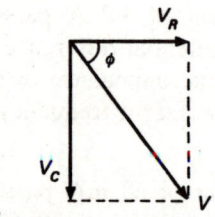

Figure 142   *Phasor for a capacitor in series with a resistor*
(a) *Resistor*
(b) *Capacitor*
(c) *Adding the voltage phasors*

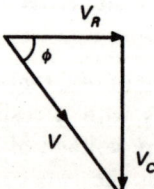

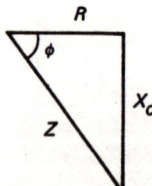

Figure 143   *The voltage triangle*   **Figure 144**   *The impedance triangle*

As $V = IZ$, $V_R = IR$ and $V_C = IX_C$ and the current is the same for the circuit as a whole, the resistor and the capacitor, then the sides of the voltage triangle can be drawn as an impedance triangle, as in Figure 144. This triangle is a useful way of summarizing the various relationships for impedance. Thus, by applying Pythagoras's theorem to the triangle gives:

$$Z^2 = X_C^2 + R^2$$

Considering the various trigonometrical relationships also gives:

$$\tan \phi = \frac{X_C}{R}$$

$$\sin \phi = \frac{X_C}{Z}$$

$$\cos \phi = \frac{R}{Z}$$

Example 12

A capacitor of 8 μF is in series with a 4 kΩ resistor. What is the impedance of the two at a frequency of 1 kHz?

$C = 8\ \mu F = 8 \times 10^{-6}$ F, $R = 4\ k\Omega = 4 \times 10^3\ \Omega$,
$f = 1\ kHz = 1 \times 10^3$ Hz

$$X_C = \frac{1}{2\pi fC}$$

$$= \frac{1}{2\pi \times 1 \times 10^3 \times 8 \times 10^{-6}} \qquad \frac{1}{Hz \times F}$$

$$= 19.9\ \Omega$$
$$Z = \sqrt{(X_C^2 + R^2)}$$
$$= \sqrt{(19.9^2 + 4000^2)}$$
$$= 4000\ \Omega$$

Note that in the above example the effect of the capacitor on the overall impedance is negligible.

Example 13

A capacitor and a series resistor together have an impedance of 2000 Ω. What will be the root mean square current through the series arrangement when the total potential difference across it is 20 V root mean square?

$$Z = 2000 \ \Omega, \ V_{rms} = 20 \ V$$

$$Z = \frac{V_{rms}}{I_{rms}}$$

Hence: $\quad I_{rms} = \dfrac{V_{rms}}{Z}$

$$= \frac{20}{2000} \qquad\qquad \frac{V}{\Omega}$$

$$= 0.01 \ A$$

Example 14

A capacitor of 16 μF is connected in series with a 500 Ω resistor and a 20 V root mean square voltage at a frequency of 50 Hz connected across the combination. Calculate the root mean square current in the circuit and the phase angle between the total potential difference and this current.

$$C = 16 \ \mu F = 16 \times 10^{-6} \ F, \ R = 500 \ \Omega, \ V_{rms} = 20 \ V, f = 50 \ Hz$$

$$X_C = \frac{1}{2\pi \, fC}$$

$$= \frac{1}{2\pi \times 50 \times 16 \times 10^{-6}} \qquad\qquad \frac{1}{Hz \times F}$$

$$= 198.9 \ \Omega$$
$$Z = \sqrt{(X_C^2 + R^2)}$$
$$= \sqrt{(198.9^2 + 500^2)}$$
$$= 538.1 \ \Omega$$

The relationship $Z = V/I$ applies where V and I are either both peak values or both root mean square values. Thus:

$$Z = \frac{V_{rms}}{I_{rms}}$$

Hence: $\quad I_{rms} = \dfrac{V_{rms}}{Z}$

$$= \frac{20}{538.1} \qquad\qquad \frac{V}{\Omega}$$

$$= 0.0372 \ A$$

Any one of the trigonometrical relationships obtained from the impedance triangle can be used to obtain φ. Thus, for example:

$$\tan \phi = \frac{X_C}{R}$$

$$= \frac{198.9}{500} \qquad \frac{\Omega}{\Omega}$$

$$\phi = 21.7°$$

This is the angle by which the potential difference lags the current.

Example 15
Draw the voltage and impedance triangles for the circuit described in Example 14.

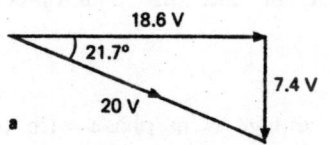

For the voltage triangle:
$$V_R = IR = 0.0372 \times 500 \qquad A \times \Omega$$
$$= 18.6 \text{ V}$$
$$V_C = IX_C = 0.0372 \times 198.9 \qquad A \times \Omega$$
$$= 7.40 \text{ V}$$

The resulting voltage triangle is shown in Figure 145(a).

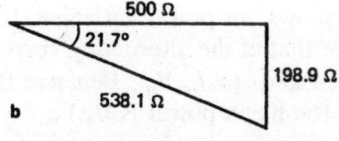

For the impedance triangle:
$$X_C = 198.9 \ \Omega \quad \text{and} \quad R = 500 \ \Omega$$

The resulting impedance triangle is shown in Figure 145(b). Both triangles are exactly the same shape, since all corresponding angles are the same. If the appropriate scale is chosen corresponding sides are the same lengths.

Figure 145 (a) *Voltage triangle* (b) *Impedance triangle*

Self-assessment questions

24 A capacitor of 2 µF is in series with a 100 Ω resistor. What is the impedance of the two at a frequency of 50 Hz?

25 A capacitor and a series resistor together have an impedance of 1.2 kΩ. What will be the root mean square current through the series arrangement when the total potential difference across it is 24 V?

26 A capacitor of 0.1 µF is connected in series with a 200 Ω resistor and a 12 V root mean square voltage at a frequency of 1 kHz. What is the root mean square current in circuit and the phase angle between the total potential difference and this current?

27 Draw the voltage and impedance triangles for the circuit described in Question 26.

28 A circuit consisting of a capacitor of 0.05 µF in series with a 500 Ω resistor takes a current of 15 mA root mean square value, at a frequency of 3 kHz. What is the total potential

difference across the two components and the phase angle between this potential difference and the current?

29 Draw the voltage and impedance triangles for the circuit described in Question 28.

Power and power factor

In any alternating current circuit, the power at any instant is the product of the instantaneous current and the instantaneous voltage, i.e.

power $P = iv$

In a purely resistive circuit, the voltage is in phase with the current. The power at any instant is the product of the instantaneous current and voltage values at the same instant and so varies with time. Figure 146(a) shows this power variation. It is a sine graph with a frequency twice that of the alternating current or voltage. The power varies from zero to I_m/V_m. Because the curve is symmetrical about $\frac{1}{2}I_mV_m$ the mean power is $\frac{1}{2}I_mV_m$.

mean power $= \frac{1}{2}I_mV_m$

Since $V_m = \sqrt{2}\,V_{\text{rms}}$ and $I_m = \sqrt{2}\,I_{\text{rms}}$,

$$\text{mean power} = \tfrac{1}{2}(\sqrt{2}\,I_{\text{rms}})\,(\sqrt{2}\,V_{\text{rms}})$$
$$= I_{\text{rms}}V_{\text{rms}} = I^2_{\text{rms}}R$$

In a purely inductive circuit the voltage leads the current by 90° (or putting it another way, the current lags the voltage by 90°). The power at any instant is again the product of the instantaneous current and voltage values at the same instant and so varies with time. Figure 146(b) shows this power variation. It is a sine graph with a frequency twice that of the alternating current or voltage. The sine curve is symmetrical about the zero power axis, with negative and positive power values. The negative values occur when a positive current is multiplied by a negative voltage or a negative current multiplied by a positive voltage. Because the power graph is symmetrical about the zero axis the mean power is zero. This can be explained as follows. During the positive power part of the power cycle, energy is being stored in the magnetic field of the inductor, whereas during the negative part of the cycle this energy is being released and put back into the circuit.

In a purely capacitive circuit the voltage lags the current by 90° (or put another way, the current leads the voltage by 90°). The power at any instant is again the product of the instantaneous current and voltage values at the same instant and so varies with time. Figure

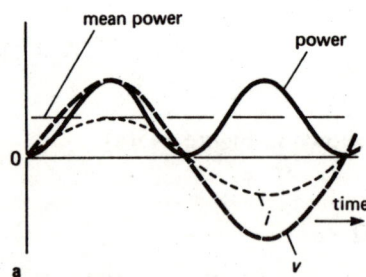

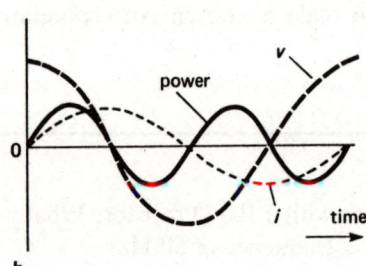

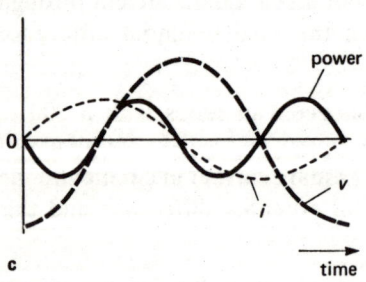

Figure 146 *Power variations with time*
(a) *Pure resistance*
(b) *Pure inductance*
(c) *Pure capacitance*

146(c) shows this power variation. It is a sine graph with a frequency twice that of the alternating current or voltage. The sine curve is symmetrical about the zero power axis, with both negative and positive power values, so the mean power is zero. This can be explained as follows. During the positive part of the power cycle energy is stored in the electric field of the capacitor, so that the capacitor becomes charged. During the negative part of the cycle the energy is released and put back into the circuit.

For a purely resistive circuit, the mean power is $I_{rms}V_{rms}$. For a circuit which has purely reactance, either inductive or capacitive, the mean power is zero.

For a circuit consisting of a resistance in series with a reactance, be it inductive or capacitive, the mean power is only that dissipated in the resistance as the reactance dissipates no power on average. Thus:

$$\text{mean power} = IV_R$$

where I is the current through the series combination and V_R is the potential difference across the resistor. But as has been earlier indicated for the voltage triangle where ϕ is the phase angle

$$\cos \phi = \frac{V_R}{\text{total r.m.s. voltage } V}$$

Hence $\boxed{\text{mean power} = IV \cos \phi}$

where I is the root mean square current through the series combination and V is the root mean square potential difference across the two components, i.e. the supply voltage. The unit of mean power is the watt.

The $\cos \phi$ factor in the above equation is the factor by which the product of the root mean square current and total potential difference must be multiplied to give the mean power. This factor is known as the power factor.

$$\boxed{\text{power factor} = \cos \phi}$$

The product of the root mean square current and the total root mean square potential difference is known as the apparent power, for which the unit is VA, rather than watt.

$$\boxed{\text{apparent power} = IV}$$

Hence

$$\boxed{\text{mean power} = \text{power factor} \times \text{apparent power}}$$

The mean power is often referred to as the true power, to distinguish it from the apparent power since it is the actual power dissipated.

Example 16
A circuit takes a root mean square current of 4 A when supplied from the 240 V a.c. mains and the power dissipated is 1000 W. What is (a) the apparent power and (b) the power factor?

$V_{rms} = 240$ V, $I_{rms} = 5$ A, true power $= 1000$ W

(a)　　　apparent power $= I_{rms} V_{rms}$
$$= 5 \times 240 \qquad\qquad \text{A} \times \text{V}$$
$$= 1200 \text{ VA}$$

(b)　　　true power $=$ power factor $\times$ apparent power
Hence　power factor $= \dfrac{\text{true power}}{\text{apparent power}}$

$$= \dfrac{1000}{1200} \qquad\qquad \dfrac{\text{W}}{\text{W}}$$

$$= 0.83$$

Example 17
A circuit consists of an inductance of 0.2 H in series with a 100 Ω resistor. What is the power dissipated when a root mean square potential difference of frequency 50 Hz and 20 V is connected across the combination?

$L = 0.2$ H, $R = 100$ Ω, $f = 50$ Hz, $V_{rms} = 20$ V
$X_L = 2\pi fL$
$$= 2\pi \times 50 \times 0.2 \qquad\qquad \text{Hz} \times \text{H}$$
$$= 62.8 \text{ Ω}$$
$Z = \sqrt{(X_L^2 + R^2)}$
$$= \sqrt{(62.8^2 + 100^2)}$$
$$= 118.1 \text{ Ω}$$
$I_{rms} = \dfrac{V_{rms}}{Z}$

$$= \dfrac{20}{118.1} \qquad\qquad \dfrac{\text{V}}{\text{Ω}}$$

$$= 0.169 \text{ A}$$

The power can be calculated by calculating the power dissipated purely in the resistance, because there is no mean power dissipation in the inductor. Hence

$$\begin{aligned} \text{power} &= I_{\text{rms}}^2 R \\ &= 0.169^2 \times 100 \qquad\qquad \text{A}^2 \times \Omega \\ &= 2.86 \text{ W} \end{aligned}$$

An alternative is to calculate the phase angle ϕ and use the relationship, power $= IV \cos \phi$.

$$\cos \phi = \frac{R}{Z}$$

$$= \frac{100}{118.1}$$

Hence power $= 0.169 \times 20 \times \dfrac{100}{118.1}$

$$= 2.86 \text{ W}$$

Example 18
A 32 µF capacitor is connected in series with a 40 Ω resistor and the combination is connected across a 240 V, 50 Hz alternating supply. What is (a) the phase angle and (b) the power dissipated?

$C = 32 \text{ µF} = 32 \times 10^{-6} \text{ F}, R = 40 \text{ }\Omega, V_{\text{rms}} = 240 \text{ V}, f = 50 \text{ Hz}$

(a) $\quad X_C = \dfrac{1}{2\pi f C}$

$$= \frac{1}{2\pi \times 50 \times 32 \times 10^{-6}} \qquad\qquad \frac{1}{\text{Hz} \times \text{F}}$$

$$\begin{aligned} &= 99.5 \text{ }\Omega \\ Z &= \sqrt{(X_C^2 + R^2)} \\ &= \sqrt{(99.5^2 + 40^2)} \\ &= 107.2 \text{ }\Omega \end{aligned}$$

$$\cos \phi = \frac{R}{Z}$$

$$= \frac{40}{107.2}$$

$$\phi = 68.1°$$

(b) $\quad I_{\text{rms}} = \dfrac{V_{\text{rms}}}{Z}$

$$= \frac{240}{107.2} \qquad \frac{V}{\Omega}$$

$$= 2.24 \text{ A}$$
$$\text{power} = I_{\text{rms}}V_{\text{rms}}\cos\phi$$
$$= 2.24 \times 240 \times \cos 68.1°$$
$$= 201 \text{ W}$$

Alternatively the power could have been calculated using power = $I^2_{\text{rms}}R$.

$$\text{power} = 2.24^2 \times 40 \qquad\qquad A^2 \times \Omega$$
$$= 201 \text{ W}$$

Self-assessment questions

30 A circuit takes a current of 2 A and dissipates a power of 400 W when supplied from the 240 V a.c. mains. What is (a) the apparent power and (b) the power factor?

31 A circuit consists of an inductance of 0.4 H in series with a 150 Ω resistor. What is the power dissipated when an alternating potential difference of root mean square value 30 V and frequency 50 Hz is connected across the combination?

32 A circuit consists of a capacitance of 8 μF in series with a 150 Ω resistor. What is the power dissipated when an alternating potential difference of root mean square value 12 V and frequency 100 Hz is connected across the combination?

33 A coil carries a current of 4 A when connected to a 240 V, 50 Hz supply and dissipates 850 W. What is the resistance and inductance of the coil?

34 A circuit consists of a 2 μF capacitor in series with a 800 Ω resistor. What is the power dissipated in each component when a 240 V, 50 Hz supply is connected across the combination?

The power triangle

If the current I in a circuit lags the voltage V by phase angle ϕ then the phasor diagram is like that shown in Figure 147(a). We can, however, resolve the voltage V into two components at right angles to each other, one of which is $V\cos\phi$ in the same direction as the current, and the other $V\sin\phi$ at right angles to the current direction. Figure 147(b) shows these components. If each of these

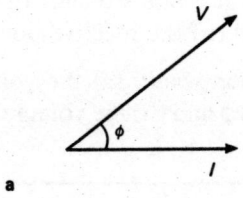

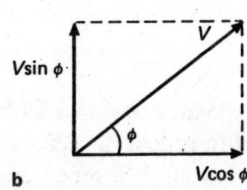

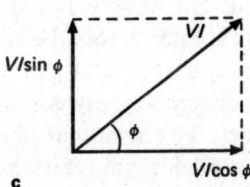

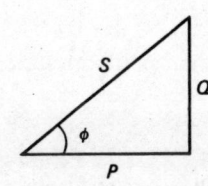

Figure 147 (a) *Phasor diagram*
(b) *Resolving V into its components*
(c) *Multiplying all sides of (b) by* I
(d) *The power triangle*

components is multiplied by I, we obtain $IV\cos\phi$ and $IV\sin\phi$, where the original phasor is IV (see Figure 147(c)). But IV is the apparent power, symbol S. The component $IV\cos\phi$ is the true power, symbol P. The other term $IV\sin\phi$ is called the reactive power, or reactive voltamps, symbol Q. Thus:

$$\begin{aligned} S &= IV \\ P &= IV\cos\phi \\ Q &= IV\sin\phi \end{aligned}$$

The reactive power is written with a unit of V Ar, where the r attached to the unit indicates that it is reactive power rather than true power.

Figure 147(c) is arrived at by the above process. It can be redrawn as a triangle (see Figure 147(d)). This triangle is known as the power triangle or power diagram.

Example 19
A transformer has an apparent power of 20 kV A at a power factor of 0.8. What is (a) the true power and (b) the reactive power?

$S = 20$ kVA $= 20 \times 10^3$ VA, power factor $= \cos\phi = 0.8$

(a) true power $P = IV\cos\phi$
   $= $ apparent power $\times$ power factor
   $= 20 \times 10^3 \times 0.8$ VA
   $= 16 \times 10^3$ W
   $= 16$ kW

(b) reactive power $Q = IV\sin\phi$
   As $\cos\phi = 0.8$, then $\phi = 36.9°$ and so $\sin\phi = 0.6$.
   $Q = $ apparent power $\times \sin\phi$
   $= 20 \times 10^3 \times 0.6$ VA
   $= 12 \times 10^3$ V Ar
   $= 12$ kV Ar

Self-assessment questions

35 Define true power, apparent power, reactive power and power factor.

36 A transformer has an apparent power of 80 kV A at a power factor of 0.8. What is (a) the true power and (b) the reactive power?

37 A load dissipates 10 kW of power and has a reactive power of 8 kVAr. What is the power factor?

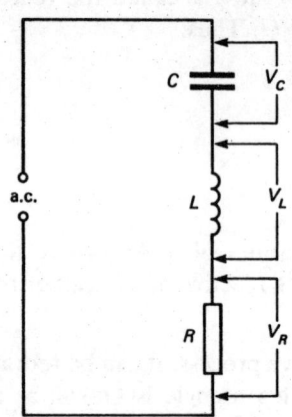

Figure 148   *A LCR circuit*

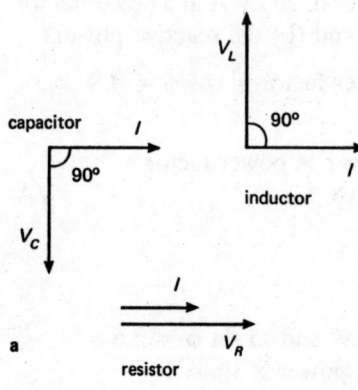

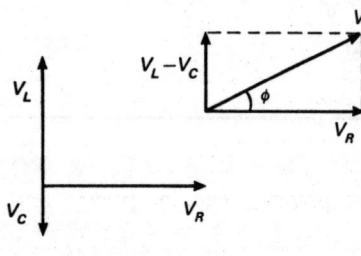

Figure 149   *The phasor diagram for a LCR circuit*
(a) *The phasors*
(b) *The phasor diagram*

38 A load dissipates a power of 6 kW at a power factor of 0.6. What is (a) the apparent power and (b) the reactive power?

39 Draw a power triangle for a situation where (a) the voltage leads the current by a phase angle of ϕ and (b) the voltage lags the current by a phase angle of ϕ.

The series LCR circuit

Figure 148 shows the basic circuit for a capacitor in series with an inductor and a resistor. The potential difference V_C across the capacitor lags the current by 90°, the potential difference across the inductor leads the current by 90°, but the potential difference across the resistor is in phase with the current. Figure 149 shows the phasors and how the total potential difference across the series combination can be obtained. Because the phasors for V_L and V_C are exactly out of phase, we can take the difference between them as being the resultant of their two phasors. This resultant phasor $(V_L - V_C)$ is then combined using the parallelogram relationship with the phasor for the resistance V_R. Hence

$$V^2 = (V_L - V_C)^2 + V_R^2$$

$$\boxed{V = \sqrt{[(V_L - V_C)^2 + V_R^2]}}$$

The potential differences are either all peak values or all root mean square values. But $V_L = IX_L$. Also $V_C = IX_C$ and $V_R = IR$, hence

$$V = \sqrt{[(IX_L - IX_C)^2 + (IR)^2]}$$
$$V = I\sqrt{[(X_L - X_C)^2 + R^2]}$$

But V/I is the circuit impedance Z, hence

$$\boxed{Z = \sqrt{[(X_L - X_C)^2 + R^2]}}$$

In the above it has been assumed that V_L is greater than V_C, and so X_L greater than X_C. If this is not the case and V_C is greater than V_L, the term in the brackets becomes $(V_C - V_L)$ and $(X_C - X_L)$.

The phase angle ϕ is given by

$$\boxed{\tan \phi = \frac{V_L - V_C}{V_R}}$$

$$\text{or} \quad \boxed{\tan \phi = \frac{X_L - X_C}{R}}$$

This is the angle by which the total voltage leads the current and is the situation arising when V_L is greater than V_C. If V_C is greater than V_L then

$$\tan \phi = \frac{V_C - V_L}{V_R}$$

$$\tan \phi = \frac{X_C - X_L}{R}$$

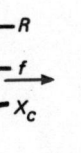

Figure 150 *Impedance in a series LCR circuit*

and ϕ is the angle by which the total voltage lags the current.

Both X_L and X_C depend on the frequency used, with $X_L = 2\pi fL$ and $X_C = 1/(2\pi fC)$. Figure 150 shows these relationships graphically. At some particular frequency however we have $X_L = X_C$. When this occurs the term $(X_L - X_C)$ is zero and so $Z = R$ and $\phi = 0$. This frequency is known as the resonant frequency f_0. At this frequency the circuit impedance is a minimum, being only equal to the circuit resistance, and so the circuit current is a maximum.

At resonance the supply voltage and the current are in phase, the circuit impedance is a minimum and the current a maximum determined solely by the circuit resistance; also, the inductive reactance equals the capacitive reactance.

This last condition means that

$$2\pi f_0 L = \frac{1}{2\pi f_0 C}$$

$$f_0 = \frac{1}{2\pi \sqrt{(LC)}}$$

The supply voltage is equal to the potential difference across the resistance V_R. Thus at the resonant frequency

$$\frac{V_L}{V} = \frac{V_L}{V_R}$$

$$= \frac{I \times \omega_0 L}{IR}$$

$$= \frac{\omega_0 L}{R}$$

The quantity V_L/V is known as the voltage magnification factor Q, thus

$$Q = \frac{\omega_0 L}{R}$$

where $\omega_0 = 2\pi f_0$. Because $\omega_0 L$ can be greater than R there can be voltage magnification with the potential difference across the inductor greater than the supply voltage.

Similarly for the potential difference across the capacitor V_C

$$Q = \frac{V_C}{V} = \frac{V_C}{V_R}$$

$$= \frac{(I/\omega_0 C)}{IR}$$

$$Q = \frac{1}{\omega_0 CR}$$

At resonance we have the situation that there is energy continually passing back and forth between the capacitor and the inductor. First, for example, we have the magnetic field in the inductor building up and energy stored in the magnetic field. Then the field collapses and the energy is passed to the capacitor as it becomes charged and stores energy in its electric field. The energy moves back and forth between the magnetic field of the inductor and the electric field of the capacitor. The only energy that has to be supplied from the supply is that necessary to compensate for the losses in the resistor.

Example 20
What is the resonant frequency for a series circuit involving a 0.2 H inductor in series with a 8 μF capacitor?

$L = 0.2$ H, $C = 8$ μF $= 8 \times 10^{-6}$ F

$$f_0 = \frac{1}{2\pi \sqrt{(LC)}}$$

$$= \frac{1}{2\pi \sqrt{(0.2 \times 8 \times 10^{-6})}}$$

$$= 126 \text{ Hz}$$

Example 21

A 2 μF capacitor is connected in series with a coil of inductance 0.1 H and resistance 50 Ω. Calculate (a) the resonant frequency, (b) the voltages at resonance, across the capacitor and the coil and (c) the circuit current, when the supply voltage is 20 V.

$C = 2\ \mu F = 2 \times 10^{-6}\ F,\ L = 0.1\ H,\ R = 50\ \Omega,\ V = 20\ V$

(a) $f_0 = \dfrac{1}{2\pi\ \sqrt{(LC)}}$

$$= \dfrac{1}{2\pi\ \sqrt{(0.1 \times 2 \times 10^{-6})}}$$

$$= 356\ Hz$$

(b) Impedance at resonance $= R = 50\ \Omega$

Hence current at resonance $= \dfrac{V}{R} = \dfrac{20}{50} = 0.4\ A$

$$V_C = IX_C$$

But $X_C = \dfrac{1}{2\pi fC}$

$$= \dfrac{1}{2\pi \times 356 \times 2 \times 10^{-6}} \qquad \dfrac{1}{Hz \times F}$$

$$= 224\ \Omega$$

Hence $V_C = 0.4 \times 224 \qquad\qquad A \times \Omega$

$$= 89.6\ V$$

(c) The coil has both resistance and inductance and so the potential difference required is that effectively across two components, a resistor and an inductor. Because the circuit is at the resonance condition we must have

$$V_L = V_C = 89.6\ V$$

The potential difference across the resistance V_R is

$$V_R = IR = \text{the supply voltage of } 20\ V$$

Thus the voltage V across the coil is given by

$$V = (V_L^2 + V_R^2)$$
$$= \sqrt{(89.6^2 + 20^2)}$$
$$= 91.8\ V$$

Example 22

A series circuit consists of 20 Ω resistance in series with a 40 Ω inductive reactance and a 60 Ω capacitive reactance. If the voltage supplied to the series arrangement is 24 V, 50 Hz, what is the current in the circuit and the potential differences across each component?

$R = 20 \; \Omega$, $X_L = 40 \; \Omega$, $X_C = 60 \; \Omega$, $V = 24$ V, $f = 50$ Hz

Note that there is no information in the question to indicate that this is a resonance condition, hence the equations for the resonance condition cannot be used. As X_L does not equal X_C, it is not reasonance.

$$\begin{aligned} Z &= \sqrt{[(X_C - X_L)^2 + R^2]} \\ &= \sqrt{[(60 - 40)^2 + 20^2]} \\ &= 28.3 \; \Omega \end{aligned}$$

But $Z = V/I$, hence

$$I = \frac{V}{Z} = \frac{24}{28.3} \qquad\qquad \frac{V}{\Omega}$$

$$= 0.848 \text{ A}$$

For the resistance

$$V_R = IR = 0.848 \times 20 = 17.0 \text{ V}$$

For the inductive reactance

$$V_L = IX_L = 0.848 \times 40 = 33.9 \text{ V}$$

For the capacitive reactance

$$V_C = IX_C = 0.848 \times 60 = 50.9 \text{ V}$$

Example 23

What is the voltage magnification across the capacitor in a series circuit of a 30 Ω resistor, a 0.06 H inductor and a 1 μF capacitor at resonance?

$R = 30 \; \Omega$, $L = 0.06$ H, $C = 1 \; \mu\text{F} = 1 \times 10^{-6}$ F

$$\omega_0 = 2\pi \, f_0 = \frac{1}{\sqrt{LC}} \qquad\qquad \frac{1}{\text{H} \times \text{F}}$$

$$= \frac{1}{\sqrt{(0.06 \times 1 \times 10^{-6})}}$$

$$= 4.08 \times 10^3 \text{ rad/s}$$

Hence $\quad Q = \dfrac{1}{\omega_0 CR}$

$$= \frac{1}{4.08 \times 10^3 \times 1 \times 10^{-6} \times 30} \qquad \frac{1}{(\text{rad/s}) \times \text{F} \times \Omega}$$

$$= 8.17$$

Self-assessment questions

40 State the conditions that exist in a series LCR circuit at resonance.

41 What is the resonant frequency for a series circuit having 0.05 H inductance and 2 µF capacitance?

42 A 16 µF capacitor is in series with an inductor of 0.5 H and a resistor of 1 kΩ. What is the resonant frequency and the voltages across the components at that frequency if the supply voltage is 10 V?

43 A coil has an inductance of 0.2 H and a resistance of 10 Ω and is in series with a 1 µF capacitor. What is the resonant frequency and the current at that frequency if the supply voltage is 12 V?

44 For the previous question what are the potential differences across the coil and the capacitor at resonance?

45 A series circuit consists of a 100 Ω resistor in series with an inductor of 0.5 H and a capacitor of 2 µF. If the voltage supply to the arrangements is 40 V, 50 Hz, what are the current in the circuit and the potential differences across each component?

46 What is the voltage magnification across the capacitor in a series circuit of a 100 Ω resistor, a 2 H inductor and a 16 µF capacitor at resonance?

47 A coil of resistance 10 Ω and an inductance 0.6 H is connected in series with a capacitor of 16 µF across a 50 V alternating supply. Calculate the frequency at which the current in the circuit is a maximum and the value of that maximum current.

8 Semiconductor diodes and transistors

After reading this chapter you should be able to:

1 Distinguish between conductors, semiconductors and insulators on the basis of their resistivities.
2 Describe the basic conduction mechanism in metals and semiconductors in terms of the charge carriers involved.
3 Explain the effect of doping on the properties of a semiconductor.
4 Describe and explain the action of a *p–n* junction.
5 Describe and explain transistor action.

Resistivity

The resistance of a length of wire depends on the type of material of which the wire is made, the length of the wire and its cross-sectional area. The longer the wire the greater its resistance, the resistance being directly proportional to the length L.

$$R \propto L$$

The greater the cross-sectional area A of the wire, the smaller its resistance, since the resistance is inversely proportional to the area.

$$R \propto \frac{1}{A}$$

Combining these two variables gives

$$R \propto \frac{L}{A}$$

Including a factor, called resistivity ρ, to take account of the differences between different materials:

$$R = \frac{\rho L}{A}$$

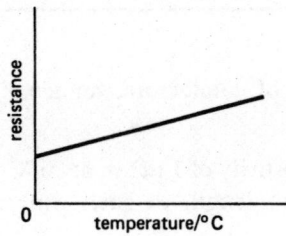

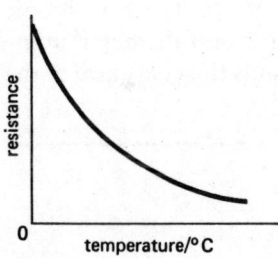

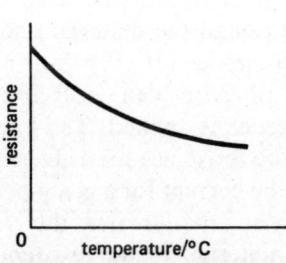

Figure 151 *The resistance scales on the graphs are not the same for the three different materials*
(a) *A metal*
(b) *A semiconductor*
(c) *An insulator*

The unit of resistivity is Ω m.

For metals the resistivities are very low, varying from about 0.016 $\mu\Omega$ m for the very good conductors such as copper and silver to 1 $\mu\Omega$ m for 'resistance' wire alloys such as nichrome. Even then nichrome can be considered to be a good conductor in comparison with many other materials.

Insulators have resistivities of the order of 10^{13} Ω m to 10^{10} Ω m. These materials have resistivities of the order of 10^{20} times those of the good conductors.

In between the resistivities of the good conductors and the insulators are a group of materials known as semiconductors. These materials, of which germanium and silicon are examples, have resistivities of the order of 1000 to 2000 Ω m.

For metals, the good conductors, the resistivity increases with temperature. Semiconductors and insulators have resistivities which decrease with increasing temperature. Thus if, for example, the resistance of a piece of material increases with an increase in temperature it must be a good conductor, a metal. If, however, the resistance decreases with an increase in temperature, it could be either a semiconductor or an insulator. Figure 151 shows how the resistances of typical conductors, semiconductors and insulators vary with temperature.

Example 1

Copper has a resistivity of 1.7×10^{-8} Ω m at 20 °C. What is the resistance of a 200 mm length of copper wire of diameter 1 mm at this temperature?

$\rho = 1.7 \times 10^{-8}$ Ω m, $L = 200$ mm $= 0.200$ m, $d = 1$ mm $= 1 \times 10^{-3}$ m

$$\text{area } A = \frac{\pi \, d^2}{4}$$

$$= \frac{\pi \times (1 \times 10^{-3})^2}{4} \qquad \text{m}^2$$

$$= 7.85 \times 10^{-7} \text{ m}^2$$

$$R = \frac{\rho L}{A}$$

$$= \frac{1.7 \times 10^{-8} \times 0.200}{7.85 \times 10^{-7}} \qquad \frac{\Omega \text{ m} \times \text{m}}{\text{m}^2}$$

$$= 4.33 \times 10^{-3} \text{ } \Omega$$

Self-assessment questions

1 State the order of resistivities of conductors, semiconductors and insulators.

2 A sample of nichrome has a resistivity of 1 μΩ m at 20 °C. What is the resistance of a 300 mm length of such wire with a diameter of 1 mm?

3 A resistance coil consists of 80 turns of nichrome wire, resistivity 1 μΩ m at 20 °C, of 1 mm diameter wound on a former of 20 mm diameter. What is the resistance of the coil at 20 °C?

Conduction in solids

For a current to flow in a material there must be 'free' charge carriers which are able to move through the material under the action of the applied potential difference. If there are no free charge carriers then no movement of charge can occur and hence no current when a potential difference is applied. The lower the resistivity of a material, the lower the resistance for a given strip of the material and hence the higher the current for a given potential difference. The higher the current, the greater the rate of movement of charge through the material. A low resistivity thus implies a greater number of free charge carriers that are able to move through the material. The term free is used because a material may have many charged particles but if they are not free to move they cannot contribute to the current.

In a metal the conduction is due to 'free' electrons. These electrons are not attached to any one atom and can be easily moved through the metal by an applied potential difference (see Figure 152). The resistivity of semiconductors is greater than that of the conductors and this indicates that there are less free charge carriers in such materials. Typically, in a material there is about

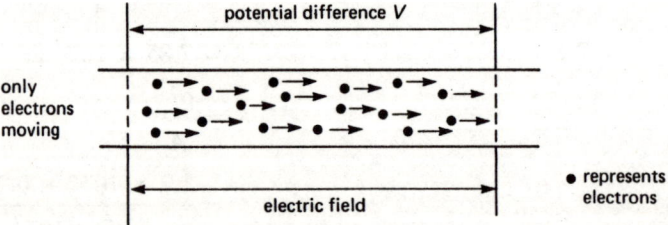

Figure 152 *The basic model of conduction in a metal*

one free charge carrier for every atom in the material, in a semiconductor there is only about one free charge carrier for every million atoms. In an insulator the resistivity is very high and the application of a potential difference to a piece of insulator results in barely any current at all. In an insulator there are hardly any free charge carriers.

In a semiconductor and an insulator, an increase in temperature results in a decrease in resistivity. This is because the increase in temperature results in an increase in the number of free charge carriers.

Self-assessment questions

4   Which of the following materials has the most, and which the fewest, free charge carriers: a metal, a semiconductor or an insulator?

5   Why does the resistivity of a semiconductor decrease when the temperature increases?

Atoms

Atoms are the basic building blocks of matter. Each atom can be considered to be made up of two parts; one is a nucleus which contains most of the mass of the atom and its positive charge, and the other part electrons which have very little mass and carry negative charge. A simple model of an atom is a small central nucleus surrounded by a cloud of electrons. Some of these electrons will be relatively close to the nucleus and some quite distant from it. Because the electrons carry a negative charge and the nucleus a positive charge there is an electric force of attraction between them. This force is responsible for the electrons remaining in the atom. Generally the total charge carried by the electrons is as big as that carried by the nucleus. Thus as one is negative charge and the other positive charge, the net charge on an atom is zero.

However, some atoms have electrons which are very distant from the nucleus. As the force of attraction between an electron and the nucleus depends on the distance between them and the greater the distance the smaller the force, these distant electrons are only very loosely held to the atom and are very easily detached. Metals have atoms in which an electron is very distant from the nucleus and so very easily detached. When the atoms of a metal are packed together into a piece of solid metal, and the metal is at room temperature, these electrons become detached from their atoms and are said to be 'free' electrons. They can be considered to be floating about between the metal atoms.

An insulator has all its electrons very tightly held to the nucleus and they are not at all easily detached. It is for this reason that there are virtually no free charge carriers in an insulator and they are very bad conductors of electricity.

Self-assessment questions

6   Why do metals have free charge carriers?

7   Why are insulators such bad conductors of electricity?

Semiconductors

Germanium and silicon are semiconductors. Their atoms have all their electrons more tightly held than is the case with metals. However, because we are considering these materials at room temperature (about 300 °C above absolute zero) enough heat energy has been received for some of the electrons to break free. However the number that can break free with this energy is very small compared with the number that are free in a metal. Thus a current can exist in a semiconductor.

If the temperature of the semiconductor is increased, more energy is supplied to the atoms and more electrons are able to break free. Hence as the temperature is increased the resistivity decreases.

When an electron breaks free from an atom it leaves that atom with a net positive charge and a vacancy for an electron. This vacancy is termed a hole. The hole can be filled by an electron jumping into it from another atom. Since each electron that moves and fills a hole creates another hole we effectively have a moving hole. In pure germanium or pure silicon we have obviously as many holes as we have free electrons. Such semiconductors are said to be intrinsic semiconductors.

When a potential difference is applied to an intrinsic semiconductor, we apply an electric field and the free electrons are therefore acted on by forces and move. They can move by hopping into and out of holes. Thus the movement of the electrons in one direction leads to an apparent movement of holes in the opposite direction. The holes behave as if they were positively charged particles. The resulting current is thus made up of two elements: electron movement in one direction and a hole movement in the opposite direction (see Figure 153).

Figure 153 *The basic model of conduction in an intrinsic semiconductor*

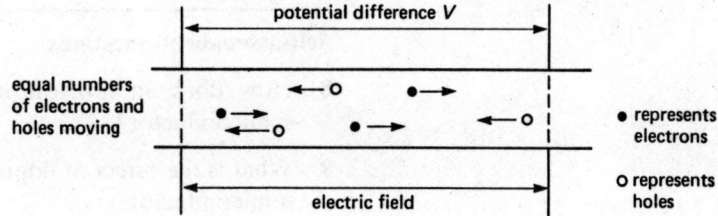

In an intrinsic semiconductor for every electron there is a corresponding hole and thus the conduction can be considered to be due in equal parts to hole and electron movement.

The introduction of 'impurities' into silicon or germanium markedly affects their resistivities. The deliberate introduction of impurities is known as 'doping'.

When small amounts of phosphorus, arsenic or antimony are added to silicon or germanium the material ends up with more free electrons than holes because these materials have easily detached electrons.

Such a material is known as n-type because the main conduction mechanism is by movement of electrons, negative charge carriers.

When small amounts of aluminium, gallium or indium are added to germanium or silicon the material ends up with more holes than free electrons.

Such a material is known as p-type, the main conduction mechanism being by the movement of holes.

Holes move in the opposite direction that electrons would in an electric field and so can be considered to behave like positive charge carriers, hence the designation as *p*-type with the *p* standing for positive (see Figure 154).

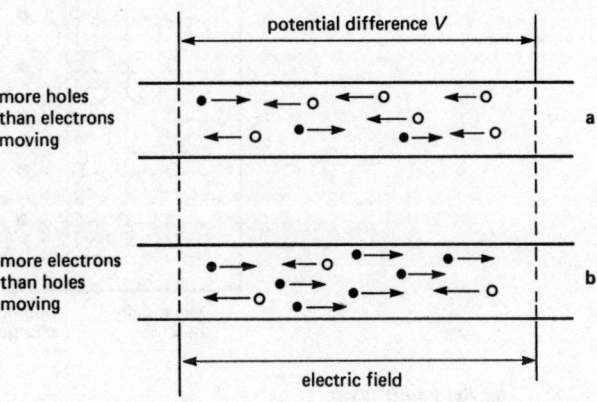

Figure 154 *The basic model of conduction in a doped semiconductor*

Self-assessment questions

8 How does an intrinsic semiconductor differ from a doped semiconductor?

9 What is the effect of doping on the conduction mechanism in a semiconductor?

The junction diode

The junction diode is a crystal of germanium or silicon in which a junction between *p*- and *n*-type material has been produced. In *p*-type material, though the net charge is zero, conduction is mainly by holes. In a *n*-type material, though again there is no net charge, conduction is mainly by electrons. When a junction between *p*- and *n*-type material is produced, the electrons in the *n*-type material close to the junction are able to drift across the junction and drop into the holes in the *p*-type material. This leaves holes behind in the *n*-type material. The result is a movement of electrons and holes across the junction so that in the narrow region on either side of the junction we have the *p*-type material gaining electrons and so becoming negatively charged and the *n*-type gaining holes (i.e., losing electrons) and so becoming positively charged. The narrow layer on either side of the junction in which this effect occurs is called the depletion layer (see Figure 155).

The movement of charge across the junction leaves the n-material positively charged and the p-material negatively charged. There is thus a potential difference across the junction. This is known as the barrier potential.

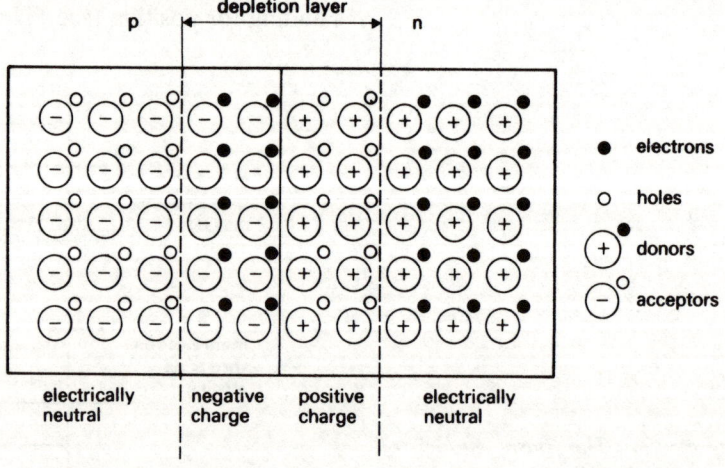

Figure 155   *The depletion layer*

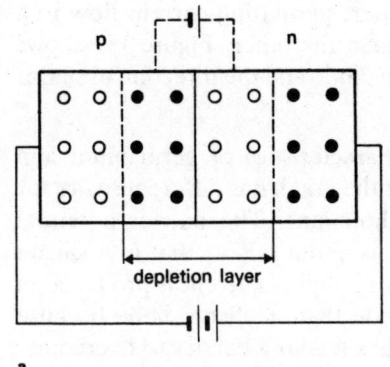

depletion layer

a

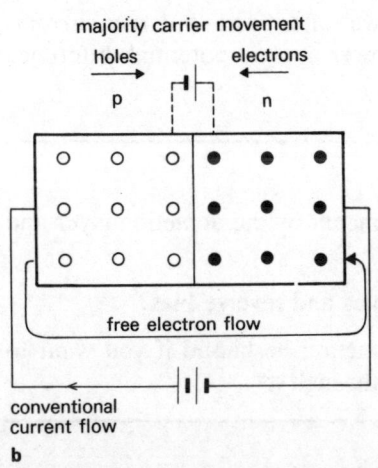

majority carrier movement

holes   electrons

free electron flow

conventional
current flow

b

Figure 156 (a) p–n *junction con-nected in the reverse bias mode* (b) p–n *junction connected in the forward bias mode*

The reason for the term barrier potential is because it opposes the further movement of charge across the junction. Thus if more electrons try to move to the *p*-material they have to overcome the repulsive force that exists because the *p*-material already has a negative charge. In a junction diode on open circuit the barrier potential, i.e., the charge distribution, builds up until it reaches an equilibrium value, which is of the order of a fraction of a volt. The equilibrium value occurs when the charge on each side of the junction reaches its maximum value. There is still movement of charge carriers across the junction but as many electrons move in one direction as move in the reverse direction, and as many holes in one direction as in the other direction.

We have the electrons from the *n*-type material trying to overcome the barrier potential and reach the holes in the *p*-type material, but also we must not forget that the *p*-type material does also contain some free electrons. In *p*-type material holes are the majority carriers but there are also electrons as minority carriers of charge. Thus some of the electrons from the *p*-type material can easily cross the junction because they are attracted by the positive charge in the *n*-type material. There are not too many of these free electrons in the *p*-type material and so there is only a small current across the junction due to these. However, such a current does mean that the charge barrier is slightly reduced and hence there is scope for some electrons to move from the *n*-type material across the boundary. A similar argument can be applied to the minority carriers in the *n*-type material. The equilibrium condition is when the minority carrier current equals the majority carrier current.

When an external battery is connected across the *p–n* junction, a current can flow when the positive of the battery is connected to the *p*-side of the junction, as the junction has a low resistance. When the positive side of the battery is connected to the *n*-side of the junction virtually no current flows, as, the junction has a high resistance (see Figure 156). When the battery is connected in such a way as to give a significant current, it is said to be applying *forward* bias; when connected in such a way as to give little current it is said to be *reverse* bias.

When the battery is connected for forward bias the potential difference it is applying across the junction cancels out the effect of the barrier potential difference, and so allows the easy movement of charge carriers across the junction. When the battery is connected for reverse bias it is increasing the barrier potential and so making it very difficult for charge carriers to move across the junction. The only current that occurs is due to the fact that this potential difference across the barrier enhances the movement of the minority charge carriers and so disturbs the equilibrium to some extent, so it is a small current.

Figure 157 *Symbol for a junction diode, the outer envelope may be omitted and also the triangle may be filled in. The arrow indicates the current direction when forward biased*

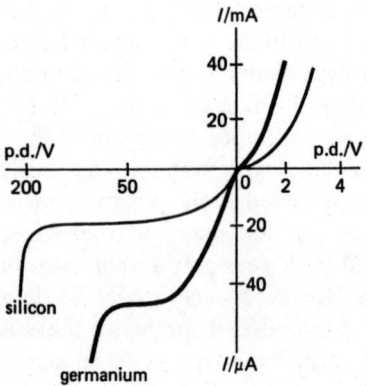

Figure 158 *The characteristics of germanium and silicon junction diodes. Note that the graph has different scales for the forward and reverse directions*

The junction thus acts as a rectifier, permitting current flow to a greater extent in one direction than the other. Figure 157 shows the circuit symbol for the junction diode and the direction of major current flow.

Figure 158 shows the typical characteristics of germanium and silicon junction diodes. Note that the graph has different scales for the forward and reverse bias directions. The barrier potential produced in a germanium diode is about 0.3 V, that in a silicon diode about 0.7 V. Because of this a higher current is produced in a forward biased germanium diode than a silicon diode because the applied potential difference has less of a barrier to overcome. If a sufficiently high reverse bias potential difference is applied, the junction diode breaks down and gives a large current. Germanium breaks down at a lower applied potential difference than silicon.

Self-assessment questions

10 For a *p–n* junction, what is meant by the depletion layer and the barrier potential?

11 What is meant by forward bias and reverse bias?

12 Which way should a *p–n* junction be biased if you want to obtain a significant current through it?

The bipolar transistor

The bipolar transistor consists of a piece of semiconducting material with a thin central region which is of the opposite type of material to that used for the two outer regions. Thus there can be a *p–n–p* transistor where the outer regions are *p*-type material and the inner narrow layer *n*-type. Similarly there can be *n–p–n* transistors where the outer regions are *n*-type material and the inner narrow layer is *p*-type. The central narrow layer is called the base, one of the outer regions is called the collector, and the other the emitter.

Figure 159 shows the basic make-up of an alloy type of bipolar transistor, in this example a *p–n–p* transistor. A slice of a crystal of *n*-type germanium, about 0.1 mm thick and 1 mm across, has pellets of indium placed on each side of it. The arrangement is then heated to about 500 °C, at which temperature the indium melts and dissolves in the germanium to form *p*-type material. The result is a narrow central region of *n*-type material with *p*-type

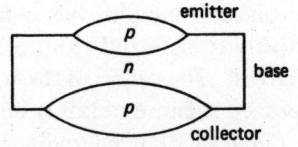

the alloyed germanium crystal slice

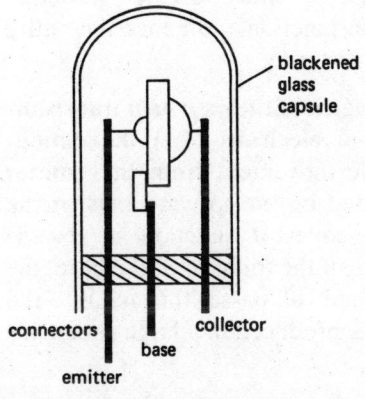

the transistor in its mounting

Figure 159 *A* p–n–p *bipolar transistor*

material on either side. There are other types and methods of constructing transistors, of which the above is just one example.

The transistor can be considered as two *p–n* junctions back-to-back with only a very small distance separating the two junctions.

To obtain transistor action the emitter-base junction has to be forward biased and the collector-base junction reverse biased.

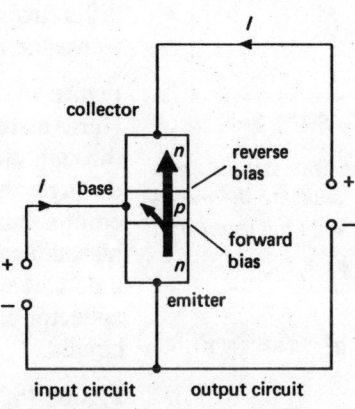

Figure 160 *Transistor action for a* n–p–n *transistor, showing the movement of electrons through the transistor*

Figure 160 shows this for a *n–p–n* transistor. There are a number of different ways of connecting transistors into circuits, but these bias conditions must occur in all of them, if transistor action is to occur.

With the circuit shown in Figure 160, known as the common emitter connection, electrons flow into the transistor at the emitter. Note that the direction of movement of electrons in a circuit is the opposite way to which the current direction is conventionally indicated. The forward bias of the emitter-base junction means that these electrons flow through into the base. Because the base is so narrow the electrons in the main end up going right through it into the collector, despite the reverse bias of the base-collector junction. The result is that there is a current in the collector circuit, despite the reverse bias.

The input signal to a transistor is applied to the emitter-base circuit. This circuit, because the emitter-base junction is forward biased, has a low resistance. We thus will have a small current from the input in a low resistance circuit. But the current, virtually unchanged, passes through the narrow base into the collector

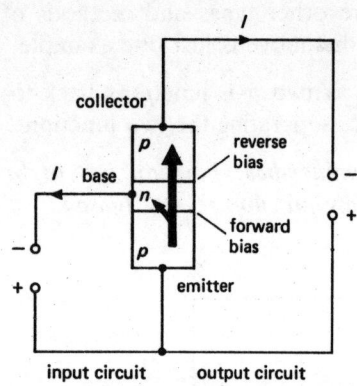

Figure 161   *Transistor action for a p–n–p transistor, showing the movement of holes through the transistor*

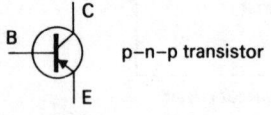

Figure 162   *Circuit symbols for transistors*

circuit. This, because of the reverse biased junction, has a high resistance. The current is thus effectively transferred from a low resistance circuit to a high resistance circuit. The origin of the term transistor is 'transfer-resistor' and was used because of this transfer of a current from a low resistance to a high resistance circuit. This transfer results in a relatively small input potential difference in the base-emitter circuit producing a small current, but this current in the collector circuit gives rise to much greater potential difference across the collector-base junction and hence the entire transistor because of the higher resistance.

Figure 161 shows the corresponding circuit for a *p–n–p* transistor. Here instead of the movement of electrons from the emitter through the base we have a hole movement from the emitter through the base. This is facilitated by the forward bias of the emitter-base junction. The base-collector junction is reverse biased for holes. However, because of the thinness of the base, the hole current from the emitter virtually all passes through into the collector circuit. Hence a current is produced in a large resistance circuit.

Figure 162 shows the symbols for *n–p–n* and *p–n–p* transistors. Note that the arrow in the *n–p–n* transistor symbol for the emitter and the arrow in the *p–n–p* transistor symbol for the emitter both represent the direction of current according to the normal convention. Such a current is in the opposite direction to that of electron flow in a circuit. The arrow can be considered to represent the direction of hole movement since holes flow in the opposite direction to electrons.

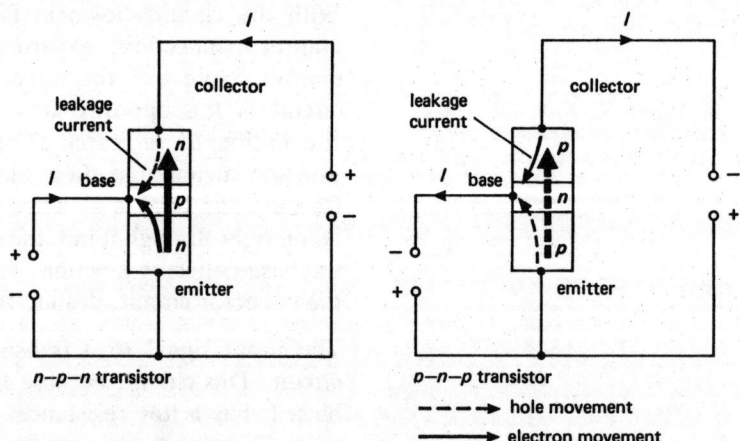

Figure 163   *Majority and minority charge carrier movements*

The charge carrier movements referred to in the above discussion of transistor action are the movements of the majority charge carriers. There will also be in all situations some movement of the minority charge carriers. Figure 163 shows the movements of both the majority and minority charge carriers for both types of transistor. The effect of minority charge carrier movement is to give a current from collector to base. This current is known as leakage current. It is small compared with the base current itself. The total base current can thus be considered to be composed of two elements, the current flowing in the base-emitter circuit and a small current leaking from the collector-base circuit.

Self-assessment questions

13 Why is it important for transistor action that the base be thin?

14 Which way should the biasing conditions be in a transistor for (a) emitter-base and (b) base-collector?

15 Draw a sketch for basic transistor action showing the correct biasing arrangements.

16 Describe the transistor action for (a) a *p–n–p* transistor and (b) a *n–p–n* transistor.

17 What is meant by the term leakage current?

Transistor circuit configurations

There are three possible ways that a transistor can be connected in a circuit. These are:

1 Common emitter mode
2 Common base mode
3 Common collector mode

Figure 164 shows these basic methods of using a transistor. The name of the mode used implies that the item mentioned, e.g. the emitter, is the common point in the input and output circuits as far as alternating signals are concerned. The common collector mode is often referred to as an emitter follower circuit. The common emitter mode of connection is probably the most widely used, although the common base mode is also widely used. The common collector mode is the least used.

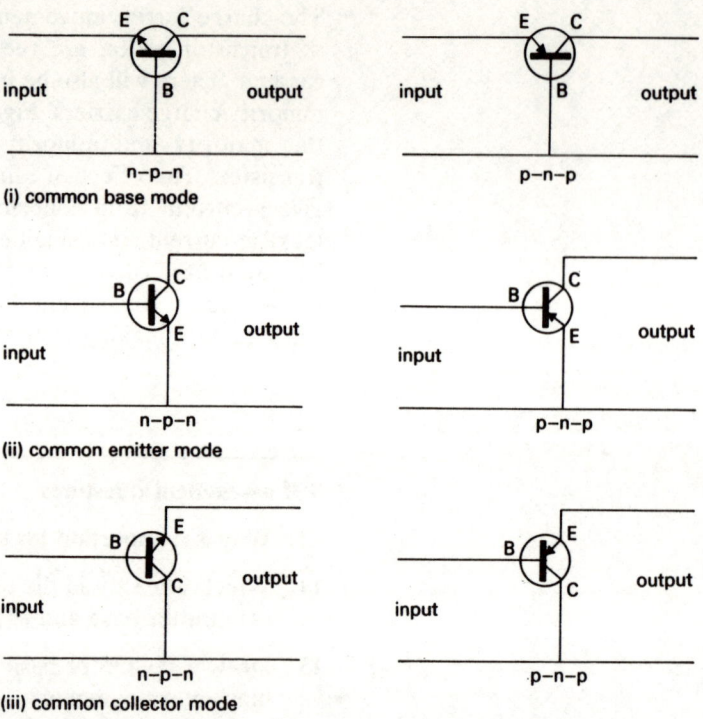

(i) common base mode

(ii) common emitter mode

(iii) common collector mode

Figure 164

In the case of the common emitter mode, the input signal gives rise to an alternating signal superimposed on the bias supplied by the battery in the input circuit. The result is a changing collector current in the higher resistance collector circuit. An output taken across the transistor, between the collector and the common input-output circuits junction, thus gives a larger voltage variation than the input voltage variation. Hence there is amplification.

Figure 165 shows how this common emitter mode of connection is used to give a practical single stage amplifier. No battery supply is included in the input circuit but a resistor is used to ensure that part of the d.c. supply in the output circuit is connected to the base. A capacitor is included in the input circuit to ensure that only the alternating signal from the source signal is fed into the base and no d.c. components from the source get through and upset the bias condition. Similarly in the output, a capacitor is included to ensure that only an alternating signal is taken from the amplifier. The circuit given is a single stage amplifier, in practice there is likely to be more than one such circuit linked together, with the output from one stage becoming the input to the next stage. In this way large amplifications can be produced.

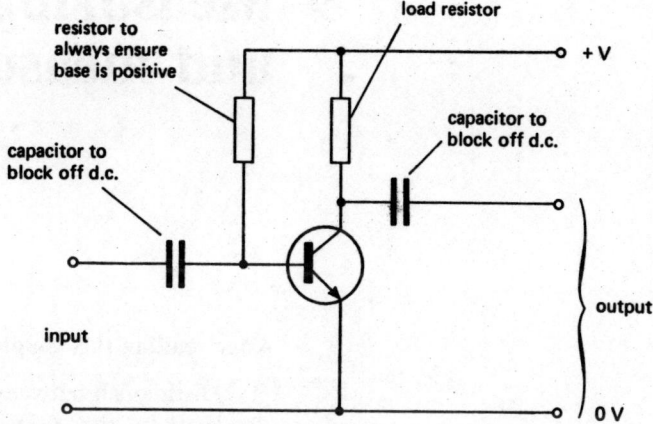

Figure 165 *Single stage transistor amplifier*

Self-assessment questions

18 State the three configurations by which a transistor can be connected in a circuit.

19 Why is the common emitter mode of connection called common emitter?

---

9 Measuring instruments and measurements

After reading this chapter you should be able to:

1 Distinguish between digital and analogue instruments.
2 Explain the principles of operation of the moving coil instrument, the limitations of the instrument and how its range can be changed.
3 Explain the principles of operation of the ohmmeter.
4 Explain the operation of multirange moving coil instruments.
5 Explain the principles of operation of the moving iron instrument, the limitations of the instrument and how its range can be changed.
6 Explain the term voltmeter sensitivity and the significance of it for voltmeter use in a circuit.
7 Explain how voltmeters and ammeters should be connected in circuits.
8 Describe common instrument errors.
9 Explain the principles of the dynamometer instrument and how it can be used as a wattmeter.
10 Explain the principles of operation of a cathode ray oscilloscope and how it can be used for potential difference and frequency measurements.
11 Explain the principles of the d.c. potentiometer and the Wheatstone bridge.

Types of measuring instrument

In general terms there are two types of measuring instrument, analogue and digital. Analogue instruments are where the deflection of a pointer across a scale gives an indication of the quantity being measured. What has happened is that the quantity being measured, perhaps a current, has been converted into another quantity, in this case a deflection, which is related to the quantity being measured. The deflection is said to be analogous to the value being measured. Digital instruments are where the reading appears in the form of a series of digits, i.e., numbers, on display tubes, a screen or perhaps printed on paper. To give an example, a watch where the time is indicated by the movement of the watch hands round a dial marked with a scale is an analogue instrument. A watch where the time is indicated as just a series of digits is a digital instrument.

Self-assessment question

1 Distinguish between analogue and digital forms of instruments.

The moving coil instrument

The moving coil instrument is a very widely used instrument. It depends for its action on a deflecting torque being experienced by a current-carrying coil when in a magnetic field produced by a permanent magnet. This instrument was discussed in Chapter 3 and a diagram of it appears as Figure 69.

The coil of wire, mounted on a rectangular frame, is suspended in the field produced by a permanent magnet. When the current passes through the coil it experiences a torque. This is due to current-carrying conductors in a magnetic field being acted on by a force. The coil rotates under the action of this torque until equilibrium is reached between this torque and a restoring torque produced by springs. The size of the deflecting torque is related to the current passed through the coil and hence, as the restoring force produced by a spring depends on the amount by which it is deflected, the equilibrium between the deflecting torque and the restoring torque occurs at a particular deflection which is related to the size of the current in the coil. By a suitable design of the magnetic field produced by the permanent magnet it is possible to obtain an instrument where the deflection of the coil is directly proportional to the current passing through the coil.

When you have used a moving coil instrument you may have noticed that when you passed a current through the instrument coil, the pointer moves across the scale, overshooting the actual current value but then swinging back and, after perhaps a few small oscillations about the actual current value, settles to a steady current value. If we only had the deflecting and restoring torques acting on the coil it would have oscillated back and forth about the current value for quite some time. It does not because there is damping. Damping provides a force which slows down the movement of the coil as it approaches the current value on the scale, so that there is perhaps only a very small overshoot of the actual value and the coil quickly comes to rest.

The damping in the moving coil instrument is generally provided by having the coil wound on an aluminium frame. This frame can be considered to be a single loop of conducting material in a magnetic field. If a loop moves in a magnetic field such that the flux linked by the loop changes, then an e.m.f. is induced in the loop. The direction of the resulting current in the loop is in such a direction that it produces a magnetic field which opposes the movement responsible for the electromagnetic induction. Thus if the coil is rotating in a particular direction the induced e.m.f. in the aluminium frame is in such a direction as to oppose the motion of the coil in that direction. This form of damping is called eddy current damping, and the current induced in the frame is called an eddy current.

A disadvantage of the moving coil instrument is that when an alternating current is supplied to the coil, instead of a direct current, the instrument generally cannot cope and just reads zero. This is because when the current through the coil is reversed the deflecting torque is reversed and the coil will endeavour to give a deflection on the other side of the zero point. With alternating current of frequency greater than about 10 Hz the coil responds too slowly to the current changes and cannot keep up with the current constantly changing from one direction to the reverse direction and just gives up and stops responding to the current. At frequencies smaller than about 10 Hz the coil may keep up to some extent with the alternating current and give a pointer movement back and forth from one side of the zero reading to the other.

The moving coil instrument can be modified to enable it to measure an alternating waveform by using it in conjunction with a rectifier. The rectifier is used to make the current fed to the instrument coil undirectional. Figure 166 shows a bridge rectifier circuit that can be used (see Chapter 6 for a discussion of the action of such a full wave rectifier). The deflecting torque for the instrument coil is proportional to the average value of the rectified current. Also the root mean square value equals the form factor multiplied by the average value (see Chapter 6 for this equation) so, if the alternating current is sinusoidal with a form factor of 1.11, the instrument can be calibrated to read root mean square values. If, however, the waveform used with the instrument has a form factor different from that of the sinusoidal waveform used to calibrate the instrument then the readings will be in error.

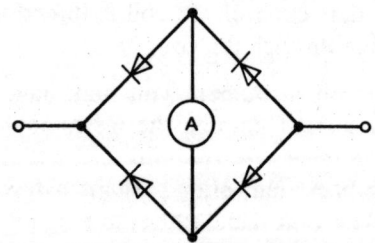

Figure 166   *A bridge rectifier with a moving coil instrument, for use with a.c.*

Rectifier instruments of the form described above can be used for frequencies up to a few kilohertz.

Self-assessment questions

2 What is the basic principle of operation for a moving coil instrument?

3 Why is damping employed with a moving coil instrument and what form does it take?

4 How can a moving coil instrument be modified so that it can be used with alternating current?

5 If a moving coil instrument with a rectifier circuit is calibrated for use with a sinusoidal waveform, what is the problem if it is used with a waveform with a different form factor?

Changing the range of a moving coil instrument

The moving coil instrument is basically an instrument for the measurement of small currents. These small currents pass through the coil of the instrument. However, the coil has a resistance and thus there is a small potential difference across it. Thus, while the instrument is dependent for its action on the current in the coil, we can also consider it to be giving an indication which is related to the potential difference across the coil. The basic instrument is often referred to as a galvanometer.

To enable the instrument to measure currents greater than that which can safely be passed through the instrument coil, a shunt is used. This is a resistor which is connected in parallel with the instrument coil and offers a path for a fixed percentage of the current. Thus one might, for instance, have a shunt which gave a path for 95 per cent of the current, leaving just 5 per cent to pass through the meter coil. The instrument can thus measure this 5 per cent, and then the instrument scale can take this into account and multiply the actual current detected by the appropriate factor to give the value of the total current.

Figure 167 shows the basic shunt arrangement. Applying Kirchhoff's first law to the currents:

$$I = I_s + I_g$$

where I is the current in the external circuit that will give a full scale deflection on the meter when a current of I_g flows through it and a current of I_s through the shunt. As the potential difference across the galvanometer must be the same as that across the shunt then

$$I_g R_g = I_s R_s$$

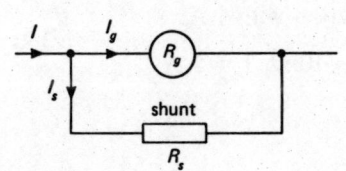

Figure 167

where R_g is the resistance of the galvanometer and R_s the resistance of the shunt.

Hence $R_s = \dfrac{I_g R_g}{I_s} = \dfrac{I_g R_g}{I - I_g}$

A moving coil meter can be used as a voltmeter, since by measuring the current through its own coil resistance it is giving an indication related to the potential difference across the coil. In order to measure higher voltages a resistance is connected in series with the instrument. This is known as a multiplier. What is happening is that the multiplier resistor takes a fraction of the total potential drop being measured and allows a constant fraction to be across the galvanometer. The instrument can then measure this fraction and then the instrument scale can take this into account and multiply the actual potential difference measured by the appropriate factor to give the value of the total potential difference.

Figure 168 shows the basic multiplier arrangement. Applying Kirchhoff's second law:

$V = V_m + V_g$
$V = I_g R_m + I_g R_g$

The current through the multiplier is the same as the current through the galvanometer I_g since they are in series. V is the voltage being measured that would give full scale deflection on the meter when V_g is the potential difference across the galvanometer and V_m that across the multiplier. R_m is the resistance of the multiplier and R_g the resistance of the galvanometer. Thus

$$R_m = \dfrac{V}{I_g} - R_g$$

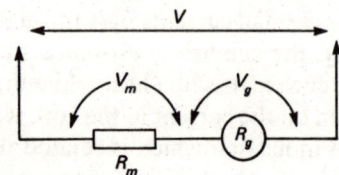

V

V_m V_g

R_m R_g

Figure 168

Example 1
A moving coil instrument gives a full scale deflection for a current of 20 mA and has a resistance of 10 Ω. Calculate the resistance of the shunt required to enable the instrument to be used as an ammeter to give a full scale deflection with 1 A.

$I_g = 20 \text{ mA} = 20 \times 10^{-3} \text{ A}, R_g = 10 \text{ Ω}, I = 1 \text{ A}$

$R_g = \dfrac{I_g R_g}{I - I_g}$

$= \dfrac{20 \times 10^{-3} \times 10}{1.0 - 20 \times 10^{-3}} \qquad \dfrac{\text{A} \times \text{Ω}}{\text{A}}$

$= 0.204 \text{ Ω}$

Example 2

A moving coil instrument gives a full scale deflection for a current of 10 mA and has a resistance of 15 Ω. Calculate the resistance of the multiplier required to enable the instrument to be used as a voltmeter to give a full scale deflection with 10 V.

$I_g = 10 \text{ mA} = 10 \times 10^{-3} \text{ A}, R_g = 15 \ \Omega, V = 10 \text{ V}$

$$R_m = \frac{V}{I_g} - R_g$$

$$= \frac{10}{10 \times 10^{-3}} - 15$$

$$= 985 \ \Omega$$

Note: Shunts have very low resistances and multipliers have very high resistances.

Example 3

Calculate the power consumed by the voltmeter, with its multiplier, in Example 2.

$I_g = 10 \text{ mA} = 10 \times 10^{-3} \text{ A}, R_g = 15 \ \Omega, R_m = 985 \ \Omega$
$P = I^2 R$

Hence for the galvanometer
$P = (10 \times 10^{-3})^2 \times 15$
   $= 1.5 \times 10^{-3} \text{ W}$

For the multiplier
$P = (10 \times 10^{-3})^2 \times 985$
   $= 0.0985 \text{ W}$
Total power consumed $= 0.0015 + 0.0985 = 0.1 \text{ W}$

An alternative to calculating the power consumed by the galvanometer and the multiplier separately is to determine the total resistance of the series combination, i.e. $985 + 15 = 1000 \ \Omega$, and then calculate the power loss in this resistance.

Self-assessment question

6 A moving coil instrument gives a full scale deflection of 10 mA and has a resistance of 15 Ω. Calculate the shunts and multipliers required for the following applications and the total power consumed in each case.
   (a)   As an ammeter giving a full scale deflection of 500 mA
   (b)   As an ammeter giving a full scale deflection of 20 A
   (c)   As a voltmeter giving a full scale deflection of 0.5 V
   (d)   As a voltmeter giving a full scale deflection of 50 V

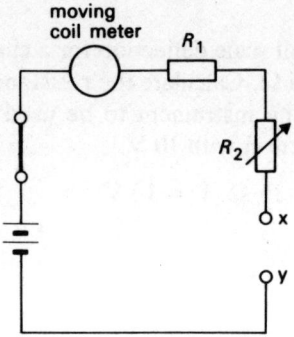

Figure 169   *The ohmmeter circuit*

The moving coil instrument as an ohmmeter

The term ohmmeter is used to describe an instrument used to measure resistance, in which the resistance is directly obtained from a pointer reading on a scale. The moving coil instrument is used as the basis of such an instrument. Figure 169 shows the basic circuit for the ohmmeter. It consists of a moving coil instrument in series with a battery, a fixed resistor and a variable resistor. With the external contacts shorted, so that the instrument is made to measure zero resistance, the value of the variable resistor is adjusted to give a full scale reading on the moving coil instrument. This reading thus represents zero resistance. Then when resistors are connected to the terminals the current reading is a measure of the resistance. The instrument scale can be calibrated in terms of resistance.

If E is the e.m.f. of the cell, R the value of the unknown resistance, R_v the resistance of the variable resistor plus the fixed resistor plus the internal resistance of the cell and R_g the resistance of the meter, then

$$E = I(R + R_v + R_g)$$

and so $$I = \frac{E}{R + R_v + R_g}$$

The current I detected by the moving coil meter is thus related to the value of R. When $R = 0$, i.e., the short circuit condition described above, then $I = E/R_v$. This resistance R_v is adjusted to give a current equal to the full scale deflection. The reason for having a fixed resistor in series with the variable resistor is to protect the instrument against too high a current being passed through it if the variable resistor becomes zero. When R is infinity, i.e., there is nothing connected between the terminals, then $I = 0$. The scale of the instrument is non-linear, ranging from 0 at one end to infinity at the other.

Example 4

An ohmmeter is to be constructed using a 1.5 V cell, of negligible internal resistance, a moving coil meter of resistance 50 Ω and a full scale deflection of 1 mA, a fixed and a variable resistor. What must be the total resistance of the fixed and variable resistances if the meter indicates a full scale deflection for zero resistance connected across the ohmmeter terminals?

$E = 1.5$ V, $R_g = 50$ Ω, $I = 1$ mA when $R = 0$
$$E = I(R + R_g + R_v)$$

Hence $$\frac{E}{I} = R + R_g + R_v$$

$$R_v = \frac{E}{I} - R - R_g$$

$$= \frac{1.5}{1 \times 10^{-3}} - 0 - 50$$

$$= 1450 \ \Omega$$

Self-assessment questions

7 What is an ohmmeter used to measure?

8 What operation must always be carried out before an ohmmeter can be used to measure the value of an unknown resistance?

9 An ohmmeter is to be constructed using a 1.5 V cell, of negligible internal resistance, a moving coil meter with a full scale deflection of 100 μA and an internal resistance of 1000 Ω, and fixed and variable resistors. What must be the value of the fixed and variable resistance when the ohmmeter is initially adjusted for zero resistance? What is the value of the external resistance which would give a current reading of (a) half the full scale reading (b) one quarter of the full scale reading?

Multirange moving coil instruments

Multirange moving coil instruments, or multimeters as they are generally referred to, enable different current ranges and different voltage ranges, both a.c. and d.c., to be obtained by operating dials on the instrument. Many also have ohmmeter ranges available. Such meters employ an array of resistors that can be switched into the circuit as shunts and others that can be switched into circuit as multipliers (see Figure 170). In addition, in order to be able to cope with a.c. a rectifying circuit will be included.

When using the multimeter to read a current or a voltage the procedure to be adopted is:

1 Switch to either d.c. or a.c. depending on the signal concerned.
2 Select the highest current range, or voltage range.
3 Only now connect the instrument to the circuit. If d.c. remember to connect the instrument in the correct polarity.
4 Now reduce the current, or voltage, range until the meter gives the largest possible reading.

When using the instrument for the measurement of resistance the procedure to be adopted is:

1 Select the resistance measurement facility by using the appropriate switches.
2 Guess the order of magnitude of the resistance being measured and switch the instrument to the best resistance range. Now adjust the set zero control so that with the instrument short circuited there is a full scale deflection and the meter indicates zero resistance.
3 Connect the resistor concerned to the instrument and take the reading.
4 If the range is wrong, repeat the zero setting operation before taking the reading on the new range.

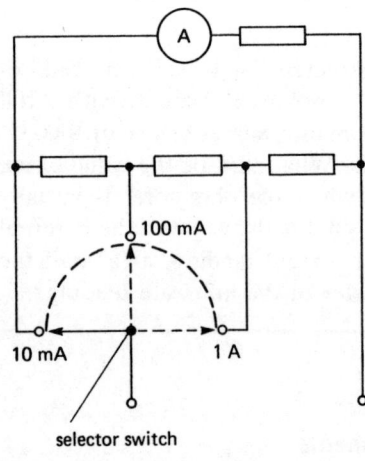

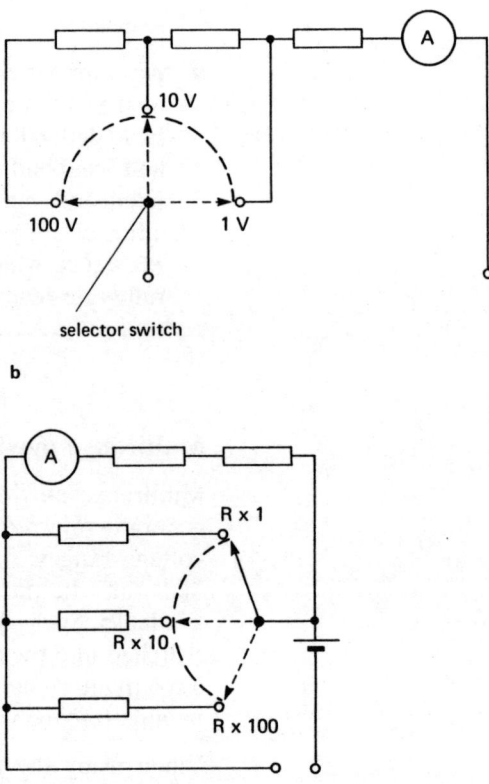

Figure 170 *Some of the types of circuit found in a multirange meter*
(a) *Current ranges*
(b) *Voltage ranges*
(c) *Resistance ranges*

Self-assessment questions

10 What steps must be taken before a multimeter is used in an electrical circuit?

11 In using a multimeter for the measurement of resistance, following the set zero adjustment, it is found that the

resistance range chosen is not the best and another range would be better. What procedure should be adopted before taking resistance readings on this other range?

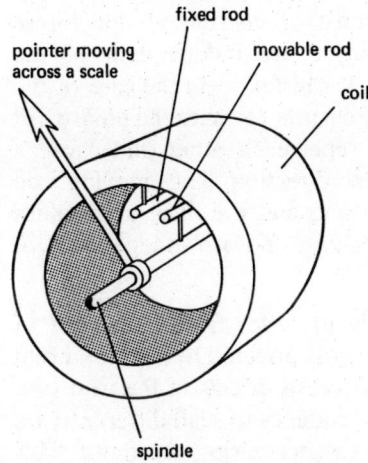

Figure 171 *The basic movement of a repulsion type moving iron instrument*

Moving iron instrument

There are two basic types of moving iron instruments, the repulsion type and the attraction type. In both types the current to be measured is passed through a fixed coil, which then produces a magnetic field. In the case of the repulsion type of instrument, the most common form of moving iron instrument, there are two rods or plates of soft iron inside the coil, one of which is fixed and the other free to move (Figure 171). These become magnetized by being in the magnetic field of the coil. Because they both become magnetized with the same polarities at similarly placed ends, and because like poles repel (see Chapter 3) a force of repulsion occurs between the two rods. The one free to move thus moves and in doing so causes a pointer to move across a scale. The force of repulsion, and hence the deflecting torque, depends on the strength of the magnets, which in turn depends on the strength of the magnetic field used to magnetize them. This depends on the size of the current in the coil. Hence the deflecting torque depends on the size of the current in the coil. This is not a linear relationship so that, for instance, doubling the current does not double the deflecting torque. As with the moving coil instruments there is a restoring torque provided by springs. The value indicated by the pointer moving across the instrument scale is the deflection that occurs when the deflecting torque is balanced by the restoring torque.

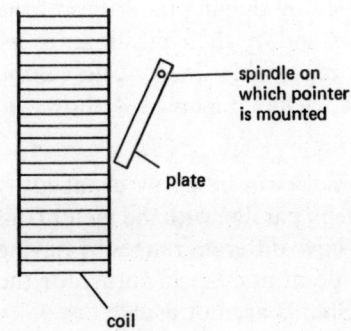

Figure 172 *End view of the basic movement of a moving iron attraction type instrument*

In the attraction type instrument, the magnetic field produced by the current through the coil attracts an iron disc towards the centre of the coil (Figure 172). Because one end of the plate is closer to the coil than the other end, it experiences a greater force. As the plate is pivoted this causes a torque and hence a deflection of a pointer across a scale. As with the repulsion type of instrument the restoring torque is provided by springs.

Damping is provided in moving iron instruments by, generally, air damping. One method of doing this is to use an air dashpot (Figure 173). The movement of the pointer across the scale causes a small plunger to be pushed into a closed cylinder containing air. The plunger compresses the air to some extent as it cannot escape fast enough past the sides of the plunger. The resulting pressure damps the motion of the pointer.

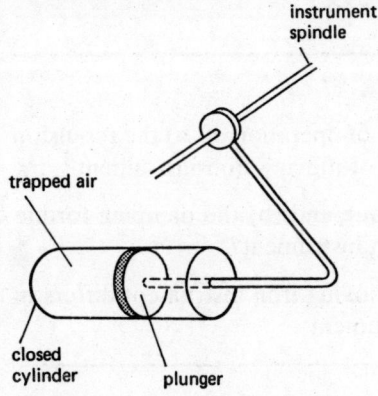

Figure 173 *Air dashpot damping*

In both the attraction and repulsion type instruments the forces responsible for the deflecting torque are independent of the current direction in the coil. Thus, for instance, in the case of the repulsion instrument, reversing the current reverses the polarity of both the iron rods but they still repel each other and cause a deflection of the pointer in the same direction. Thus moving iron instruments can be used for both a.c. and d.c. The a.c. values indicated are root mean square values, and are independent of form factor.

Moving iron instruments are liable to suffer from a number of errors. They are, for instance, markedly affected by stray external magnetic fields. Also, owing to hysteresis effects in the iron (see Chapter 4), the instruments have a tendency to read differently on increasing current as opposed to decreasing current. This hysteresis effect also affects their performance at different frequencies.

Moving iron instruments have a non-linear scale. That means that doubling the current does not double the deflection of the pointer. The form of the scale can be changed by shaping the forms of the iron pieces used in the instruments and by their positioning. By careful design of such elements a reasonably linear scale can be achieved over a major part of the scale. Figure 174 shows an example of such a scale.

The range of a moving coil instrument can be easily changed by connecting a shunt, i.e., a resistance in parallel with the meter coil. Moving iron instruments generally have different ranges by having a number of coils each, with a different number of turns, for the production of the magnetic field. Shunts are not usually used.

Moving iron instruments used for the measurement of voltage usually have high resistance coils consisting of a large number of turns of fine wire. A voltage multiplier, i.e., a resistance in series with the instrument coil, is often used.

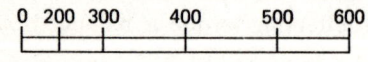

Figure 174 *Typical type of scale with a moving iron instrument*

Self-assessment questions

12 What are the basic principles of operation of (a) the repulsion and (b) the attraction forms of moving iron instrument?

13 How is (a) the restoring torque and (b) the damping torque produced with a moving iron instrument?

14 State two ways by which a moving iron instrument differs in use from a moving coil instrument.

Voltmeter sensitivity

The sensitivity of a voltmeter is the total resistance of the voltmeter, including any multipliers in use, divided by the voltage required on that range to give full scale deflection. It has the units ohms per volt.

$$\text{sensitivity} = \frac{\text{total resistance of voltmeter}}{\text{voltage to give full scale deflection}}$$

The higher the sensitivity, i.e., the greater the resistance per volt, the better the instrument is. A high quality instrument may have a sensitivity of the order of 20 000 Ω/V.

The reason that the voltmeter with the higher resistance per volt is better is that the instrument is placed in parallel with the component across which the potential difference is to be measured, and ideally should take no current from the circuit so that the current through the component remains unchanged. If significant current is taken, the potential difference being measured is significantly changed by connecting the voltmeter into the circuit. In other words, the potential difference indicated by the meter is not the value it originally was. The higher the resistance of the voltmeter, the less the current taken by the instrument and so the less the effect on the potential difference being measured. Hence the higher the sensitivity, the less the use of the instrument will affect the potential difference being measured.

Example 5

A moving coil instrument has a resistance of 15 Ω. When used as a voltmeter for a full scale reading of 10 V, a multiplier of 985 Ω is used, what is the sensitivity of the voltmeter on this range?

$R_g = 15\ \Omega$, $V = 10$ V, $R_m = 985\ \Omega$

Total voltmeter resistance = 15 + 985 = 1000 Ω

$$\text{sensitivity} = \frac{\text{total resistance of voltmeter}}{\text{voltage to give full scale deflection}}$$

$$= \frac{1000}{10} \qquad \frac{\Omega}{V}$$

$$= 100\ \Omega/V$$

Self-assessment questions

15 What is meant by the sensitivity of a voltmeter?

16 Which is the better voltmeter, one with a low sensitivity or one with a high sensitivity? Explain your answer.

17 A moving coil instrument has a resistance of 10 Ω. When used as a voltmeter for a full scale reading of 20 V, a multiplier of 9990 Ω is used. What is the sensitivity of the voltmeter on this range?

Using voltmeters and ammeters

Voltmeters are used to determine the potential difference across a component and so are connected in parallel with the component. Ammeters are used to measure the current through a component and so are placed in series with the component. In general voltmeters have high resistances so that when they are connected in parallel with a component they draw little current and do not significantly affect the current through the component and hence the potential difference being measured. Ammeters have low resistances so that including them in the circuit does not significantly affect the current in the circuit.

When the component for which the current and potential difference and current are being measured has a high resistance (perhaps approaching that of the voltmeter) the obvious method of connecting the instruments (as in Figure 175) can lead to errors. In such a situation the current taken by the voltmeter can be significant in comparison with the current taken by the component and thus the current measured is not the current through the component but the sum of the currents through the component and the voltmeter. The current reading is thus in error. Figure 176 shows an alternative method of connecting the instruments. The current measured is now just the current through the component. However, this time the voltage reading is in error, since the reading is that for the potential difference across both the component and the ammeter. However, if the ammeter resistance is very small compared with that of the component, the error can be very small. Where the component has a high resistance, this method of connecting the instruments is to be preferred.

Self-assessment questions

18 Which instrument would you expect to have the higher resistance – a voltmeter or an ammeter?

19 When a moving coil voltmeter and a moving coil ammeter are used to measure the potential difference across a high resistance resistor and the current through it, what is the best method of connecting the meters to do this with the minimum error?

Instrument errors

As has been indicated in Figures 175 and 176 there can be errors in the readings given by instruments which are due to the technique adopted for making the measurements. Such errors are known as systematic errors.

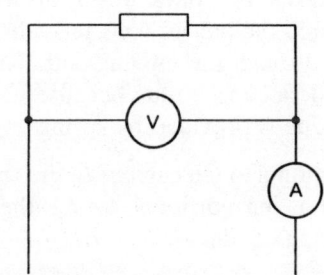

Figure 175

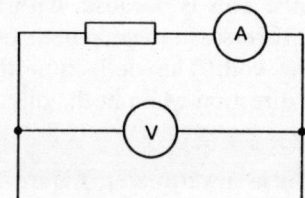

Figure 176

Another type of error can occur when the observer takes the readings from the instruments. If you change the position at which you view the pointer in an instrument, you might obtain a different value because the angle at which you view the pointer on the scale can affect the reading. To eliminate this effect, so that everybody looks at the pointer and the scale at the same angle each time they use the instrument, meters often have a small mirror placed behind the pointer. The reading is taken when the pointer is lined up with its image in the mirror. This can only happen when the observer is looking along a line at right angles to the mirror, so this technique ensures that the same angle is used for each observation. This type of error which can arise from observation is called an observational error.

Another type of error is a calibration error. An instrument may be quoted by the manufacturer as having an accuracy of ± 2 per cent of full scale deflection. This means that if that instrument is used on a range with a full scale deflection of, say, 1 A, any reading may be as much as 0.02 A greater or less than the true reading would be. Thus all that can be said is that the current value indicated by such an instrument can be in error by plus or minus 0.02 A, just due to the overall accuracy with which the instrument has been calibrated.

In some cases the error due to the calibration may not be constant over the range of readings indicated by the instrument. If an instrument is to be used for accurate readings it will need to be calibrated over its entire range of readings.

Self-assessment questions

20 Name three types of error that can occur when readings are taken from instruments.

21 A manufacturer's data sheet for a voltmeter states that it has an accuracy of ± 5 per cent of full scale deflection. What does this signify?

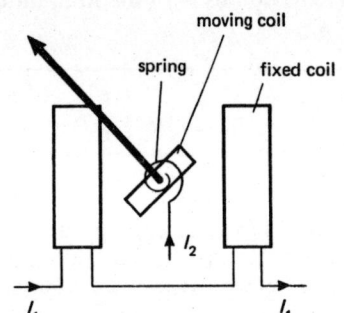

Figure 177   *The basic movement of a dynamometer instrument*

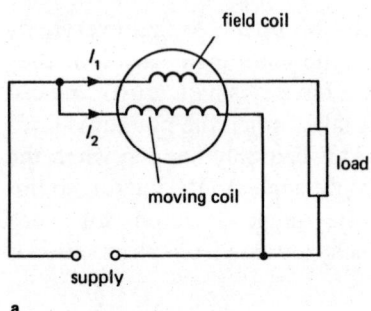

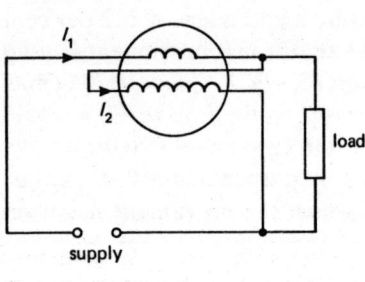

Figure 178   *Wattmeter connections (a) Conventional connection, the power loss in the field coil is included with that of the load. Most suitable for high load resistance (b) Alternative connection, the power loss in the moving coil is included with that of the load most suitable for low load resistance*

Dynamometer instruments

A dynamometer instrument is rather like a moving coil instrument, but with the main difference that the magnetic field is provided by a current flowing in a pair of coils rather than a permanent magnet. Figure 177 shows the basic form of the instrument. A current I_1 flows through the pair of coils providing the magnetic field and a current I_2 through the moving coil. The current-carrying coil experiences a deflecting torque because it is in a magnetic field. A restoring torque is provided by springs.

Because the magnetic field is proportional to the current I_1 and the force on the current-carrying coil is proportional to I_2, then (remember force on a current-carrying conductor $F = BIL$):

deflecting torque $\propto I_1 I_2$

As well as giving a response with d.c. currents the instrument can also be used with alternating currents. This is because, with the alternating current, when the current reverses in the magnetic field coils it also reverses in the moving coil. The deflection thus becomes independent of the current direction when both coils are supplied with alternating current.

An application of the dynamometer is as a wattmeter. Figure 178 shows how the instrument should be connected for such a use. The magnetic field coils carry the load current of the circuit concerned and the coils consist of a few turns of large diameter wire, hence a low resistance. The moving coil consists of many turns of fine wire and is connected so that the current through it is proportional to the potential difference across the load. Thus

$$I_1 = \text{load current}$$
$$I_2 \propto \text{potential difference across load}$$
and so $I_1 I_2 \propto$ load current × load potential difference
$$\propto \text{average power consumed by load}$$

Thus the deflection of the pointer with such a form of dynamometer is proportional to the average power consumed by the load, hence the term wattmeter.

Self-assessment questions

22 How does a dynamometer differ from a moving coil instrument?

23 Explain how the two sets of coils are connected in a wattmeter when it is used for the measurement of power.

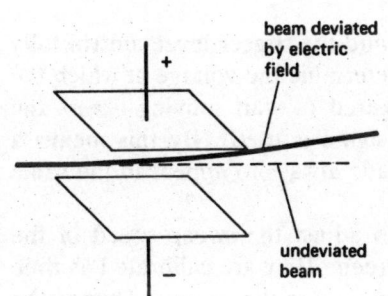

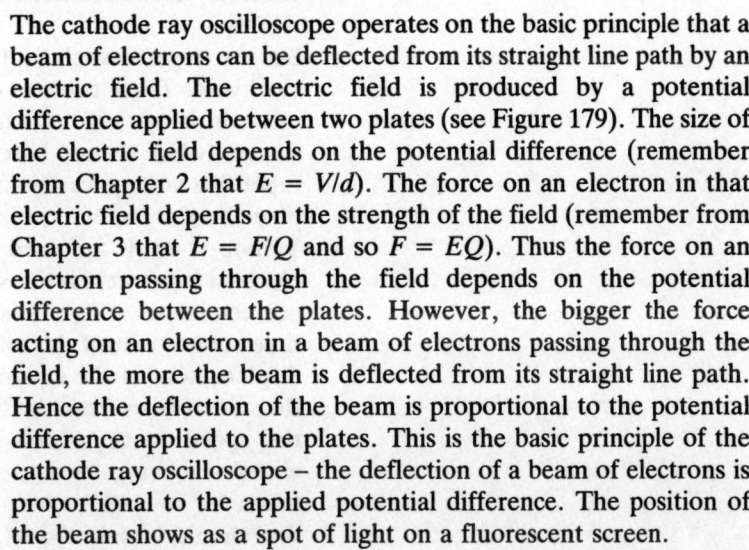

beam deviated by electric field

undeviated beam

Figure 179

The cathode ray oscilloscope

The cathode ray oscilloscope operates on the basic principle that a beam of electrons can be deflected from its straight line path by an electric field. The electric field is produced by a potential difference applied between two plates (see Figure 179). The size of the electric field depends on the potential difference (remember from Chapter 2 that $E = V/d$). The force on an electron in that electric field depends on the strength of the field (remember from Chapter 3 that $E = F/Q$ and so $F = EQ$). Thus the force on an electron passing through the field depends on the potential difference between the plates. However, the bigger the force acting on an electron in a beam of electrons passing through the field, the more the beam is deflected from its straight line path. Hence the deflection of the beam is proportional to the potential difference applied to the plates. This is the basic principle of the cathode ray oscilloscope – the deflection of a beam of electrons is proportional to the applied potential difference. The position of the beam shows as a spot of light on a fluorescent screen.

In an oscilloscope there are two sets of deflection plates, one producing a deflection in the X-direction (the horizontal direction) and the other producing a deflection in the Y-direction (the vertical direction). These plates are known as the X- and Y-plates.

The controls of an oscilloscope involve more than just those concerned with the X- and Y-plates, there are also controls concerned with initial adjustments to the electron beam. There is a control to adjust the brightness of the spot or trace made on the fluorescent screen where the electrons hit it. There is also a control to adjust the focus of the beam of the electrons: an out of focus beam gives a blurred image on the screen, while an in focus beam is a sharp image.

The following, with brief explanatory notes, is a typical sequence of operations that might be followed in setting up an oscilloscope. Before plugging in:

1 Turn brightness control to the off position. The brightness control is generally also the on-off switch for the instrument.
2 Set the focus control to its midway position. There is then a reasonable chance that the beam will not be too far out of focus to miss being seen.

3   Set the X- and Y-shift controls to their midway positions. These controls are used to shift the entire beam either sideways or upwards. They supply a steady internal d.c. supply to the plates. The midway positions give you a reasonable chance that when you switch on, the trace caused by the electrons hitting the screen will be on screen.

4   Set the X- and Y-gain controls relatively low. These are the controls which determine the amplification given to the input signals.

5   Set the trigger switch to + and the trigger level control fully clockwise. These controls determine the voltage at which the electron beam will be triggered to start moving across the screen (when a time base signal is used). By this means a repetitive voltage can be made always to appear in the same place on the screen.

6   Timing controls are used to adjust the sweep speed of the electron beam across the screen. They are calibrated as time per centimetre of movement across the screen. Thus if the time base control is set to 1 ms/cm, every centimetre displacement of the electron beam across the X-direction of the screen takes 1 ms. Set this, initially, to about the middle of the speed range, unless you know the frequency of the waveform being used as input and can calculate the time taken for say one or two cycles and hence the time control that should be selected to give, say, one or two cycles on the screen.

7   Plug in. Switch on by rotating the brightness control so that a bright trace is produced on the screen.

8   Focus to obtain a clear, sharp, trace. It is often useful to reduce the brightness to a relatively low level during this operation.

9   Apply the input voltage to the Y-input, the X-input is assumed to be connected internally to the time base signal. Unless you deliberately switch the time base off it will automatically give a signal in the X-direction. Adjust the Y-gain and the time base control so that the signal occupies a reasonable percentage of the screen and an adequate number of cycles are visible.

10 Adjust the trigger control, possibly switching it to automatic, so that the trace remains stationary on the screen (assuming a repetitive input waveform).

The above are fairly general instructions on setting up an oscilloscope and obtaining a trace of a potential difference waveform on the screen. The following are more specific notes on how certain measurements can be made.

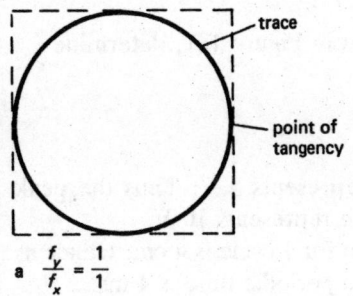

a $\dfrac{f_y}{f_x} = \dfrac{1}{1}$

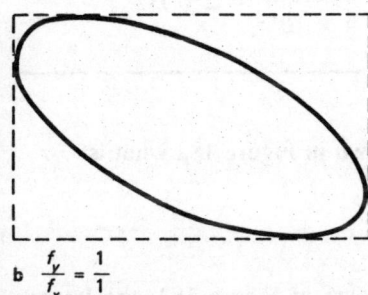

b $\dfrac{f_y}{f_x} = \dfrac{1}{1}$

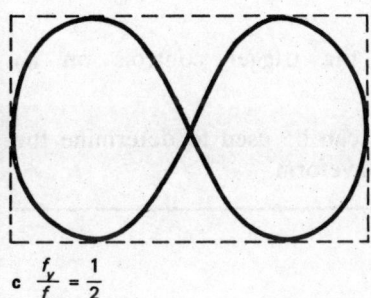

c $\dfrac{f_y}{f_x} = \dfrac{1}{2}$

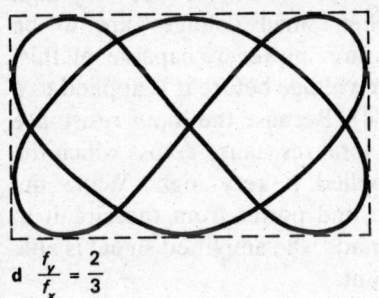

d $\dfrac{f_y}{f_x} = \dfrac{2}{3}$

Figure 180 *Examples of Lissajou's figures*

For the measurement of a direct potential difference, apply the d.c. signal to the Y-input. If the time base is left running, the signal seen on the screen will be a horizontal line. If the time base is switched off there will be a spot of light on the screen. The problem with trying to determine the value of the potential difference is that there is no automatic zero level with an oscilloscope. You have to find the signal position for zero potential difference input. So you need to take two readings of the line or spot position, one with no input and one with the potential difference input. Determine from the graticule across the screen the vertical distance between these two positions. Then read the gain value from the calibrated time base and use this to convert the vertical distance into volts. Note that this method can only be used if the Y-gain is set at a calibrated position. If not you will have to calibrate the screen by connecting known d.c. voltages to the Y-input.

For the measurement of an alternating potential difference, apply the a.c. signal to the Y-input. If the time base is switched off you will obtain a vertical line on the screen. The length of this vertical line is twice the maximum value of the alternating waveform. With the appropriate time base and the correct trigger setting the waveform will be shown on the screen. The peak-to-peak value can be measured, and also the time taken for a complete cycle. This is obtained by measuring the distance along the X-direction that one cycle takes and then using the calibration given for the time scale to convert it into time. The frequency is the reciprocal of the periodic time.

A more accurate method of measuring frequency is to compare the unknown frequency with a known frequency. The unknown signal is applied to the Y-plates and the known frequency to the X-plates, the time base having been switched off. Usually the known frequency is taken from a signal generator and adjusted so that a stable picture appears on the oscilloscope screen. The figure that appears on the screen is known as Lissajou's figure. Figure 180 gives some examples. The unknown frequency is obtained by considering a rectangle drawn round the trace. The ratio of the frequency on the Y-plates to the frequency applied to the X-plates is the ratio of the number of points of tangency on a horizontal side of the rectangle to the number of points of tangency on a vertical side.

An important point to remember in the use of an oscilloscope, is that it has a very high impedance. It can thus be considered to be a very high sensitivity voltmeter.

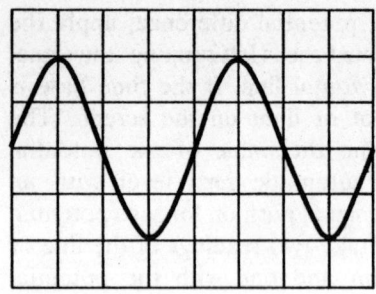

Figure 181 *Each square has a 1 cm side. The time base is 1 ms cm^{-1} and the Y gain is 5 V cm^{-1}*

Example 6

For the oscilloscope trace illustrated in Figure 181, determine:
(a) The peak potential difference
(b) The periodic time
(c) The frequency

(a) For the Y-scale, each 1 cm represents 5 V. Thus the peak value which is a length of 2 cm represents 10 V.
(b) The distance in the X-direction for 1 cycle is 4 cm. Hence as 1 cm represents 1 ms, then the periodic time is 4 ms.
(c) frequency $= \dfrac{1}{\text{periodic time}} = \dfrac{1}{4 \times 10^{-3}} = 250$ Hz

Self-assessment questions

24 For the oscilloscope trace shown in Figure 182 what is:
(a) The peak-to-peak value
(b) The periodic time
(c) The frequency?

25 For the oscilloscope trace shown in Figure 183, the known frequency applied to the X-plates was 100 Hz. What is the frequency on the Y-plates?

26 What is the function of the trigger controls on an oscilloscope?

27 Explain how an oscilloscope can be used to determine the frequency of an alternative waveform.

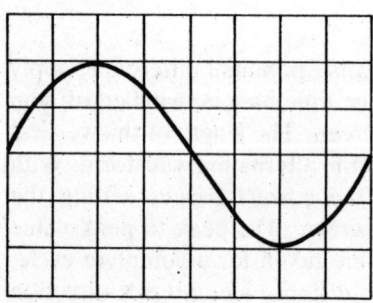

Figure 182 *Each square has a 1 cm side. The time base is 10 ms cm^{-1} and the Y gain is 10 V cm^{-1}*

Electronic voltmeters

Moving coil and moving iron instruments cannot give very high meter resistances, especially when small voltages are to be measured. Electronic voltmeters are, however, capable of this. Amplifiers are used to amplify the voltage before it is applied to a moving coil meter (see Figure 184). Because the input resistance of the voltmeter is very high, the total resistance across which the input potential difference is applied is very high. While the amplifier takes negligible current and power from the circuit in which the measurement is being made, the amplified signal is able to operate a moving coil instrument.

In using an instrument which has part of its circuit connected to the ground (or earth), care has to be taken when the instrument is used in a circuit which also has a ground connection. If both the instrument and the circuit ground connections are not essentially

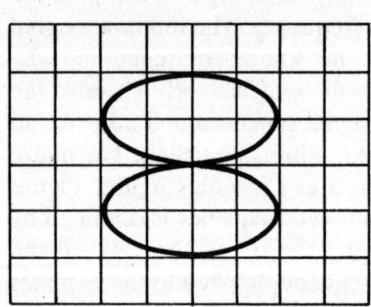

Figure 183

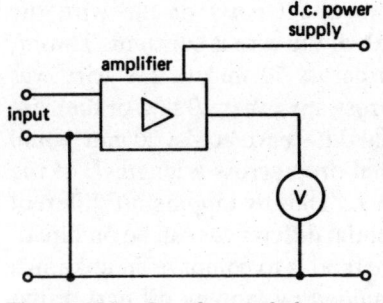

Figure 184 *One form of electronic voltmeter*

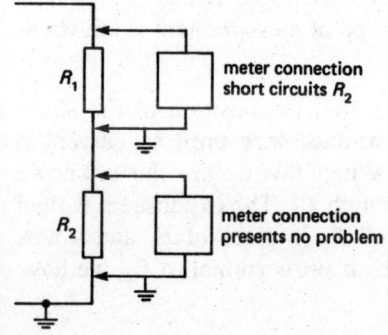

Figure 185

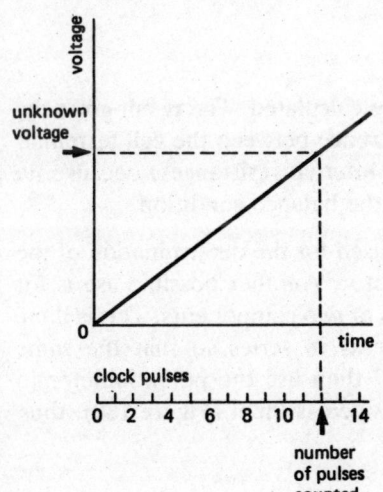

Figure 186 *A ramp voltage used with a digital voltmeter*

to the same point in the circuit, a short circuit can occur across components and the voltages being measured can be drastically changed. Figure 185 illustrates this. Electronic voltmeters are available with grounded terminals or floating and ungrounded. For the voltmeter illustrated in Figure 184 the lower line would probably be grounded. The point also applies to the use of cathode ray oscilloscopes as voltmeters in circuits.

Self-assessment question

27 What is the main advantage of an electronic voltmeter when compared with a moving coil meter?

Digital voltmeters

These instruments involve an analogue to digital converter, because the input signal is in a continuous form and has to be converted into the digital form which is non-continuous. One method that is used for digital voltmeters involves counting from an accurate electronic clock system the number of pulses that occur from the time a ramp voltage starts rising from zero until it equals the voltage being measured. A ramp voltage is a voltage that rises at a rate proportional to time. The pulse count by this method is thus proportional to the unknown voltage. Figure 186 illustrates this method.

The number of digits employed in a digital voltmeter determines the size of the increments of voltage that can be detected. This is called the resolution of the instrument. Thus a three digit display for a 999 V range instrument would mean that it is only capable of detecting an increment of 1 V. It can, for example, detect a change from 57 to 58 V but no smaller change, such as 57 V to 57.3 V.

Self-assessment question

28 A digital voltmeter has four digits. What is the smallest voltage increment that can be detected on an instrument reading up to 9.999 V?

The potentiometer

The d.c. potentiometer is a circuit used for the comparison of direct voltages. Figure 187 shows the basic circuit. An e.m.f., from the cell E, is applied across a length of uniform resistance wire.

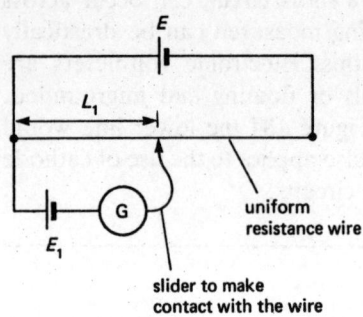

Figure 187   *The d.c. potentiometer*

Because the current is the same at all parts of the wire the potential difference per unit length of the wire is constant. Thus if, for instance, the potential drop across 10 mm of the wire was 0.2 V, then the potential drop across any other 10 mm of the wire would be 0.2 V. Also the potential difference across 20 mm would be $2 \times 0.2 = 0.4$ V. The potential drop across a length L of the wire is proportional to the length L. Thus by tapping off different lengths of the wire, different potential differences can be obtained. The basis of the potentiometer method is to compare an unknown potential difference with that obtained by tapping off part of the wire and to vary the potential difference tapped off until it cancels out that of the unknown potential difference. When this occurs the galvanometer in the potentiometer circuit shows no current.

The sole function of the galvanometer in the circuit is to detect when there is zero current. This type of measurement is known as a null measurement technique.

With the circuit shown in Figure 187, the position of the slider would be adjusted along the resistance wire until no current is detected by the galvanometer. When this occurs the unknown e.m.f. E_1 is proportional to the length L_1. The experiment is then repeated with a known e.m.f. cell E_2 in place of E_1 and a new balance length of L_2 found. As E_2 is proportional to L_2, we have

$$E_1 \propto L_1$$
$$E_2 \propto L_2$$

Hence $\boxed{\dfrac{E_1}{E_2} = \dfrac{L_1}{L_2}}$

Hence the unknown e.m.f. can be calculated. The result gives the e.m.f. and not the potential difference between the cell terminals (see Chapter 1 for discussion of internal resistances) because no current is taken from the cell at the balance condition.

The same form of circuit can be used for the determination of the potential difference across a resistor. Another possible use is for the comparision of the resistances of two components. This is done by connecting the two components in series so that the same current passes through them and then use the potentiometer to compare the potential differences across them (Figure 188), thus

$$V_1 = IR_1$$
$$V_2 = IR_2$$

Hence $\dfrac{V_1}{V_2} = \dfrac{R_1}{R_2}$

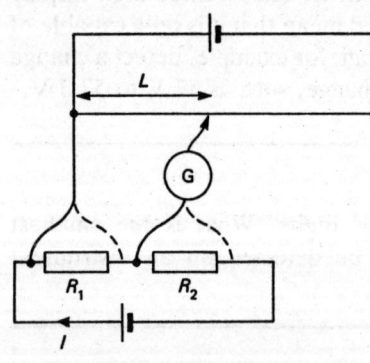

Figure 188   *Comparison of resistances using the d.c. potentiometer. The length is measured for zero current for the solid line connections and then for the dotted line connections*

But the potentiometer enables the two potential differences to be compared.

$$\frac{V_1}{V_2} = \frac{L_1}{L_2}$$

Hence $\quad \dfrac{R_1}{R_2} = \dfrac{L_1}{L_2}$

An important point of note in any form of the potentiometer circuit is that the cell supplying the e.m.f. across the full length of the wire must be connected in such a way that its positive terminal is connected to the same end of the wire as is the positive terminal of the cell being measured, or the positive side of a potential difference. Without this the two potential differences applied across the length L of the wire cannot cancel out and give zero current through the galvanometer.

Example 7

A d.c. potentiometer is calibrated against a cell of e.m.f. 1.019 V, and a balance with zero current obtained with a length of 450 mm. When the unknown e.m.f. is used the balance length is found to be 754 mm. What is the unknown e.m.f.?

$E_2 = 1.019$ V, $L_2 = 450$ mm, $L_1 = 754$ mm

$$\frac{E_1}{E_2} = \frac{L_1}{L_2}$$

$$\frac{E_1}{1.019} = \frac{754}{450}$$

$$E_1 = 1.707 \text{ V}$$

Example 8

A d.c. potentiometer is used to compare the resistances of two resistors through which the same current flows. For one resistor a balance length is obtained with 340 mm of potentiometer wire, for the other resistor the balance is with 825 mm. How do the two resistances compare?

$$\frac{R_1}{R_2} = \frac{L_1}{L_2}$$

$$= \frac{340}{825}$$

$$= 0.412$$

Example 9

The potential difference across a 10 Ω resistor is found to give a balance with a d.c. potentiometer when the length of wire is 360 mm. When a standard cell of e.m.f. 1.019 V is used the balance length is 624 mm. What is the current through the 10 Ω resistor?

For the 10 Ω resistor, if the current through it is I then the potential difference across it is $V_1 = IR = 10\,I$. Hence, as

$$\frac{V_1}{V_2} = \frac{L_1}{L_2}$$

$$\frac{10\,I}{1.019} = \frac{360}{624}$$

$$I = 0.0589 \text{ A}$$

Self-assessment questions

29   A d.c. potentiometer is calibrated against a cell of e.m.f. 1.205 V, and a balance with zero current obtained with a length of 286 mm. When the unknown e.m.f. is used the balance length is found to be 852 mm. What is the unknown e.m.f.?

30   A d.c. potentiometer is used to compare the resistances of two resistors through which the same current flows. For one resistor, its potential difference is found to require 287 mm of wire for balance, for the other resistor 682 mm. If this last resistor has a value of 6 Ω, what is the value of the other resistance?

31   The potential difference across a 5 Ω resistor is found to give a balance with a d.c. potentiometer of 658 mm. When a standard cell of e.m.f. 1.019 V is used the balance length becomes 456 mm. What is the current through the 5 Ω resistor?

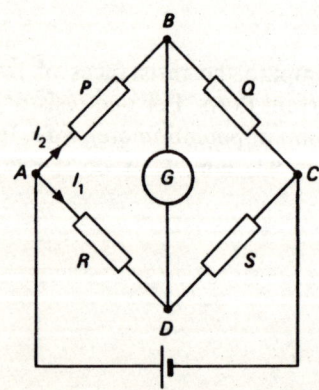

Figure 189   *The Wheatstone bridge*

The Wheatstone bridge

Figure 189 shows the basic Wheatstone bridge circuit. It is a null measurement method in that the values of the four resistors are adjusted until there is no current detected by the galvanometer. When there is no current in the galvanometer then there can be no potential difference between B and D. This means that B and D are at the same potential. Hence for the arms AB and AD of the bridge, because A is a common point we must have the potential difference across P equal to the potential difference across R. Hence

$$I_1P = I_2R$$

Similarly the potential difference across Q must equal the potential difference across S. As no current flows through the galvanometer the current through Q must be the same as the current through P, also the current through S must be the same as the current through R. Hence

$$I_1Q = I_2S$$

Dividing the first equation above by the second equation gives

$$\boxed{\frac{P}{Q} = \frac{R}{S}}$$

This is the relationship between the resistances when there is no current through the galvanometer.

In practice if P is the unknown resistance, the ratio of the resistances R/S might be chosen and Q adjusted to give the balance condition. The e.m.f. of the cell plays no part in the calculation, all that needs to be known are the ratio of two resistors and the value of a third.

Example 10
Four resistors are connected in a Wheatstone bridge, as in Figure 189. If when there is no current through the galvanometer the ratio R/S is 100 and Q is 15.6 Ω, what is the value of P?

$R/S = 100$, $Q = 15.6$ Ω

$$\frac{P}{Q} = \frac{R}{S}$$

$$P = 100 \times 15.6 \qquad\qquad \Omega$$
$$= 1560 \ \Omega$$

Self-assessment questions

32 Four resistors are connected in a Wheatstone bridge, as in Figure 189. If when there is no current through the galvanometer $R = 50$ Ω, $S = 400$ Ω and $Q = 10.1$ Ω, what is the value of P?

33 Four resistors are connected in a Wheatstone bridge, as in Figure 189. If when there is no current through the galvanometer R/S has the ratio 100 and Q is 5.65 Ω, what is the value of P?

Solutions to self-assessment questions

Chapter 1

1 $R = V/I = 1.2/1 = 1.2\ \Omega$

2 $V = IR = 0.5 \times 20 = 10\ \text{V}$

3 $I = V/R = 0.2/40 = 0.005$ A or 5×10^{-3} A

4 $R = V/R = 3.6/0.12 = 30\ \Omega$

5 To one end the voltmeter and the ammeter, to the other end the other terminal of the voltmeter and a wire from the 20 Ω resistor.

6 $I = V/R = 2.3/10 = 0.23$ A

7 The same as through the 10 Ω resistor, i.e. 0.23 A

8 At a constant temperature the potential difference is proportional to the current.

9 Also trebled.

10 (a) See Figure 190 (b) Yes (c) $R = V/I = 2/0.08 = 25\ \Omega$

11 (a) See Figure 191 (b) Only up to a current of about 0.1 A
(c) $R = V/I = 1.2/0.100 = 12\ \Omega$

12 (a) 50 mA (b) 0.2 kA (c) 400 μA (d) 0.02 A = 20×10^{-3} A
(e) 3000 A or 3×10^3 A (f) 300 μA (g) 2 kV (h) 100 μV
(i) 0.015 V = 1.5×10^{-2} V (j) 0.003 V = 3×10^{-3} V (k) 2.5 kΩ
(l) 0.14 MΩ (m) 4 000 000 Ω or 4×10^6 Ω (n) 5000 Ω or 5×10^3 Ω

13 $R = V/I = 0.6/(20 \times 10^{-3}) = 30\ \Omega$

14 $V = IR = (20 \times 10^{-6}) \times (0.5 \times 10^6) = 10$ V

15 $I = V/R = 4/(2 \times 10^{-3}) = 2000\ \Omega = 2\ \text{k}\Omega$

16 $20 + 40 = 60\ \Omega$

17 $1 + 5 + 10 = 16\ \text{k}\Omega$

18 (a) $V_1 = IR_1 = 20 \times 10^{-3} \times 10 = 0.20$ V;
$V_2 = IR_2 = 20 \times 10^{-3} \times 20 = 0.4$ V;
$V_3 = IR_3 = 20 \times 10^{-3} \times 100 = 2$ V
(b) $10 + 20 + 100 = 130\ \Omega$
(c) $0.2 + 0.4 + 2 = 2.6$ V;
or $V = IR = 20 \times 10^{-3} \times 130 = 2.6$ V

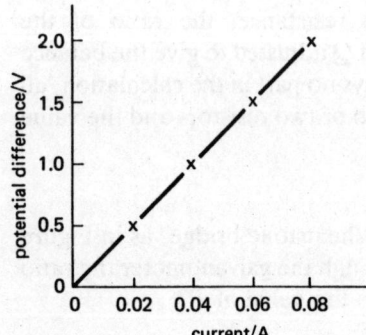

Figure 190

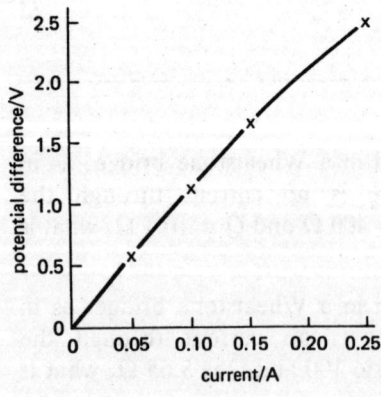

Figure 191

19 One 50 Ω resistor with three 10 Ω resistors, all in series.

20 $1 = 12 \times (R/100)$, hence $R = 8.3$ Ω. This solution used the equation given in the text. An alternative is to argue that $\frac{1}{12}$ of the voltage means $\frac{1}{12}$ of the resistance, i.e. $100 \times (1/12) = 8.3$ Ω.

21 Using the equation given in the text, $2 = 6 \times 50/(50 + R)$, hence $R = 100$ Ω. You could argue that $\frac{2}{6}$ of the voltage is across 50 Ω therefore the $\frac{4}{6}$ should be across 100 Ω (twice the voltage, twice the resistance).

22 $1/R = (1/20) + (1/10) = (3/20)$. Hence $R = 20/3 = 6\frac{2}{3}$ Ω.

23 $1/R = 3/12$, hence $R = 4$ Ω.

24 (a) $1/R = (1/100) + (1/40) = (4 + 10)/400$,
        hence $R = 400/14 = 28.6$ Ω
     (b) $V = IR = (10 \times 10^{-3}) \times 28.6 = 0.286$ V
     (c) $V = I_1R_1 = 0.286 = I_1 \times 100$, hence $I_1 = 0.002\ 86$ A = 2.86 mA;
        $I_2 = 10 - 2.86 = 7.14$ mA.

25 (a) $1/R = (1/2) + (1/1) = (3/2)$, hence $R = 2/3$ kΩ.
        $V = (20 \times 10^{-3}) \times (\frac{2}{3} \times 10^3) = 13.3$ V
     (b) $V = I_1R_1 = 13.3 = I_1 \times 2000$, hence $I_1 = 6.67$ mA.
        $I_2 = 20 - 6.67 = 13.33$ mA.

26 $1/5 = (1/R) + (1/50)$, hence $1/R = (1/5) - (1/50) = (10 - 1)/50$ and so $R = 50/9 = 5.56$ Ω.

27 (a) For parallel resistors $1/R = (1/20) + (1/40) = (3/40)$,
        hence $R = 40/3 = 13.3$ Ω. Hence total resistance $= 10 + 13.3 + 30 = 53.3$ Ω.
     (b) The 5 Ω and 20 Ω combine to give 25 Ω. The parallel resistors
        give $1/R = (1/25) + (1/30) = (30 + 25)/(25 \times 30)$, hence $R = (25 \times 30)/55 = 13.6$ Ω. This combined with the series 10 Ω resistor gives a total of 23.6 Ω.
     (c) The 20 Ω and 40 Ω in parallel give $1/R = (1/20) + (1/40) = 3/40$,
        hence $R = 40/3 = 13.3$ Ω. Combining this with the 10 Ω resistor gives 23.3 Ω. This in parallel with the 50 Ω resistor gives $1/R = (1/23.3) + (1/50) = (50 + 23.3)/(23.3 \times 50) = 0.0637$ and hence $R = 15.9$ Ω.

28 (a) For the 5 Ω and the 40 Ω the equivalent is 45 Ω, for the other
        parallel arm the 20 Ω and the 30 Ω give an equivalent of 50 Ω. The parallel arrangement thus has a resistance given by $1/R = (1/45) + (1/50) = (50 + 45)/(45 \times 50)$, hence $R = (45 \times 50)/95 = 23.7$ Ω. This combined with the series resistor of 10 Ω gives a total resistance of 33.7 Ω.
     (b) Total circuit current $= V/R = 6/33.7 = 0.178$ A. This is the
        current through the 10 Ω resistor.
     (c) Potential difference across 10 Ω resistor $= IR = 0.178 \times 10 = 1.78$ V. Hence potential difference across parallel arrangement $= 6 - 1.78 = 4.22$ V. Current through 20 Ω $= V/R = 4.22/(20 + 30) = 0.084$ A.

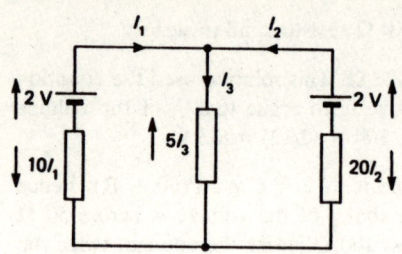

Figure 192

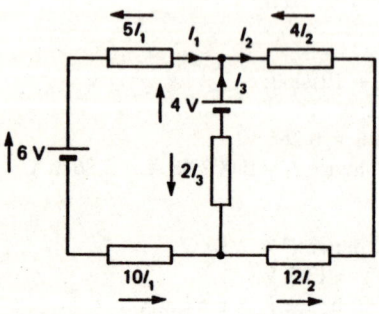

Figure 193

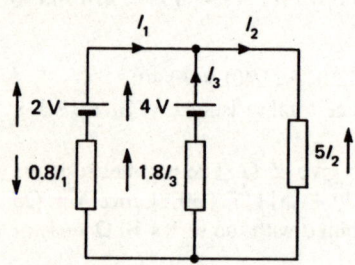

Figure 194

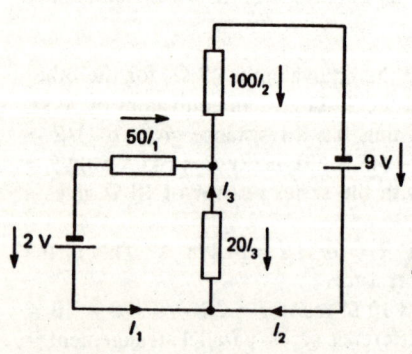

Figure 195

29 (a) Parallel circuit consists of 20 Ω in parallel with $(5 + 40 + 30) = 75$ Ω. Hence resistance given by $1/R = (1/20) + (1/75) = (75 + 20)/(20 \times 75)$ and so $R = 15.8$ Ω. This combined with the 50 Ω in series gives 65.8 Ω for total resistance. $I = V/R = 9/65.8 = 0.137$ A.

(b) Potential difference across 50 Ω resistor $= IR = 0.137 \times 50 = 6.85$ V. Therefore potential difference across $AB = 9 - 6.85 = 2.15$ V.

30 $V = E - Ir$, hence $r = (E - V)/I = 0.7/0.8 = 0.875$ Ω

31 $V = E - Ir = 2.1 - (0.2 \times 0.5) = 2.0$ V

32 See Figure 192. Applying first law $I_3 = I_1 + I_2$. Applying second law $2 - 5(I_1 + I_2) - 10I_1 = 0$ and also $2 - 5(I_1 + I_2) - 20I_2 = 0$. You may arrive at different equations if you took different current directions for your version of the circuit. The answer in the end should, however, be the same. Hence $2 - 15I_1 - 5I_2 = 0$ and $2 - 5I_1 - 25I_2 = 0$. Hence $I_1 = (2 - 5I_2)/15 = (2 - 25I_2)/5$ and so $I_2 = 0.057$ A (57 mA).

33 See Figure 193. Applying the first law $I_3 = I_1 - I_2$. Applying the second law $6 - 5I_1 - 4 + 2(I_1 - I_2) - 10I_1 = 0$ and also $4 - 4I_2 - 12I_2 - 2(I_1 - I_2) = 0$. Hence $I_1 = (2 - 2I_2)/13 = (4 - 14I_2)/2$ and $I_2 = 0.270$ A (270 mA).

34 See Figure 194. Applying the first law $I_3 = I_1 - I_2$. Applying the second law $2 - 4 - 1.8(I_1 - I_2) - 0.8 I_1 = 0$ and also $4.0 - 5I_2 + 1.8(I_1 - I_2) = 0$. Hence $I_1 = (-2 - 1.8I_2)/2.6 = (6.8I_2 - 4)/1.8$ and so $I_2 = 2.01$ A.

35 See Figure 195. Applying the first law $I_3 = I_1 + I_2$. Applying the second law $2 - 20I_3 - 50I_1 = 0$ and also $9 - 20I_3 - 100I_2 = 0$. As we require I_3 the second equation is rewritten with I_2 replaced. Hence $9 - 20I_3 - 100(I_3 - I_1) = 0$. Hence $I_1 = (2 - 20I_3)/50 = (120I_3 - 9)/20$ and so $I_3 = 0.079$ A. Thus $V = IR = 0.079 \times 20 = 1.58$ V.

36 $P = I^2R = 0.50^2 \times 100 = 25$ W

37 $P = V^2/R = 15^2/50 = 4.5$ W

38 Total resistance $= 20 + 50 = 70$ Ω, hence current $I = V/R = 4/70 = 0.057$ A.

(a) $P = I^2R = 0.057^2 \times 20 = 0.065$ W;
$P = I^2R = 0.057^2 \times 10 = 0.032$ W

(b) Total power $= 0.065 + 0.032 = 0.097$ W

39 $1/R = (1/20) + (1/10) = 3/20$, hence $R = 20/3$ Ω. $V = IR = 0.50 \times (20/3) = 10/3$ V.

(a) $P = V^2/R = (10/3)^2/20 = 0.55$ W;
$P = V^2/R = (10/3)^2/10 = 1.11$ W

(b) Total power $= 0.55 + 1.11 = 1.66$ W

40 (a) $P = I^2R = 0.15^2 \times 100 = 0.23$ W. For parallel combination $1/R = (1/40) + (1/100) = 14/400$, hence $R = 28.57$ Ω. $V = IR = 0.15 \times 28.57 = 4.286$ V. Hence $P = V^2/R = 4.286^2/40 = 0.46$ W. $P = V^2/R = 4.286^2/100 = 0.18$ W

(b) Total power $= 0.23 + 0.46 + 0.18 = 0.87$ W

Chapter 2

1 They must either both have a positive charge or both a negative charge.

2 (a) Polythene (b) Copper

3 Current = rate of movement of charge, hence for a current of 4 A there must be a movement of 4 C per second.

4 Current = rate of movement of charge = $100 \times 10^{-3}/40 = 2.5 \times 10^{-3}$ A (250 mA).

5 Current = rate of movement of charge, hence a current of 400 mA means 400×10^{-3} C/s. For 2 minutes (120 s), charge = $400 \times 10^{-3} \times 120 = 48$ C.

6 Electric field strength = force per unit charge.

7 $E = F/Q = 0.1/(0.4 \times 10^{-6}) = 0.25 \times 10^{6}$ N

8 $E = F/Q$ hence $F = EQ = 20 \times 300 \times 10^{-6} = 6 \times 10^{-3}$ N

9 The direction of the electric field.

10 (a) Electric potential at a point is the electric potential energy per unit charge at that point.
   (b) Electrical potential difference between two points is the difference in potential energy per unit charge between the two points. Electrical potential difference is the work done in moving unit charge between the two points.

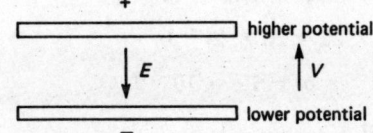

Figure 196

11 The gradient of a hill is the change in height of the hill with horizontal distance. The potential gradient is the change in potential with distance. $E = V/d$ = potential gradient. You may find this equation written as $E = -V/d = -$ potential gradient. This is because the potential increases when we move in a direction opposite to the direction of the electric field, see Figure 196.

12 $E = V/d = 20/(2 \times 10^{-3}) = 10\,000$ V/m

13 $W = QV$

14 $W = (20 \times 10^{-6}) \times 12 = 2.4 \times 10^{-4}$ J

15 $W = QV = 0.5 \times 2 = 1$ J

16 Capacitance is defined by the equation $C = Q/V$. One way of expressing this is: a charge Q given to an isolated conductor raises its potential by V, the ratio Q/V being known as the capacitance.

17 Farad (F). This is a very large unit and most capacitors have values of microfarads or less.

18 $Q = CV = (16 \times 10^{-6}) \times 12 = 1.92 \times 10^{-4}$ C. The answer is $+ 1.92 \times 10^{-4}$ C and $- 1.92 \times 10^{-4}$ C.

19 $Q = CV = (500 \times 10^{-6}) \times 12 = 6 \times 10^{-3}$ C. When charged one plate will have $+ 6 \times 10^{-3}$ C and the other $- 6 \times 10^{-3}$ C. These are the amounts of charge that will flow off each plate during the discharge.

20 Plate area, separation of the plates, the material between the plates.

21 Which was the positive and which the negative terminal.

22 (a) $C = \varepsilon_r\varepsilon_o A/d = 1 \times 8.85 \times 10^{-12} \times 0.02/(0.50 \times 10^{-3}) = 3.5 \times 10^{-10}$ F

(b) $C = 7 \times 3.5 \times 10^{-10} = 2.5 \times 10^{-9}$ F

23 $C = \varepsilon_r\varepsilon_o A/d = 3.2 \times 8.85 \times 10^{-12} \times (60 \times 10^{-4})/(0.10 \times 10^{-3}) = 1.7 \times 10^{-9}$ F

24 $C = \varepsilon_r\varepsilon_o A/d$, hence $A = Cd/\varepsilon_r\varepsilon_o = 0.1 \times 10^{-9} \times 1 \times 10^{-3}/(1000 \times 8.85 \times 10^{-12}) = 1.1 \times 10^{-5}$ m^2.

25 $C = \varepsilon_r\varepsilon_o(n - 1)A/d = 4.0 \times 8.85 \times 10^{-12} \times 8 \times 100 \times 10^{-4}/(0.2 \times 10^{-3}) = 1.4 \times 10^{-8}$ F

26 $C = \varepsilon_r\varepsilon_o(n - 1)A/d$, hence $(n - 1) = Cd/\varepsilon_r\varepsilon_o A = 8.85 \times 10^{-9} \times 0.5 \times 10^{-3}/(2.5 \times 8.85 \times 10^{-12} \times 250 \times 10^{-4}) = 8$. Hence $n = 9$.

27 $E = V/d = 100/(0.2 \times 10^{-3}) = 5 \times 10^5$ V/m.

28 The dielectric strength is the potential gradient in the dielectric at which its insulation breaks down.

29 $14 \times 0.2 = 7$ kV

30 These are the maximum voltages at which the capacitor should be used if breakdown is not to occur. However a margin of safety is allowed in the values quoted.

31 Energy $= \frac{1}{2}CV^2 = \frac{1}{2} \times 16 \times 10^{-6} \times 12^2 = 1.15 \times 10^{-3}$ J

32 Energy $= \frac{1}{2}QV = \frac{1}{2} \times 2 \times 10^{-6} \times 20 = 20 \times 10^{-6}$ J

33 Energy $= \frac{1}{2}CV^2 = \frac{1}{2} \times 100 \times 10^{-6} \times 6^2 = 1.8 \times 10^{-3}$ J

34 $C = C_1 + C_2 = 20 + 10 = 30$ µF

35 $C = C_1 + C_2 + C_3 = 1 + 2 + 8 = 11$ µF

36 $(1/C) = (1/C_1) + (1/C_2) = (1/2) + (1/4) = (3/4)$, hence $C = 4/3$ µF

37 $(1/C) = (1/C_1) + (1/C_2) + (1/C_3) = (1/4) + (1/12) + (1/16) = (19/48)$, hence $C = 48/19 = 2.53$ µF

38 $(1/C) = (1/C_1) + (1/C_2) = (1/10) + (1/20) = (3/20)$, hence $C = (20/3)$ µF. Charge $Q = CV = (20/3) \times 12 = 80$ µC. Hence $V_1 = Q/C_1 = 80/10 = 8$ V and $V_2 = 12 - 8 = 4$ V.

39 (a) $C = C_1 + C_2 = 2 + 1 = 3$ µF.
$1/C = (1/C_1) + (1/C_2) = (1/3) + (1/0.1) = (1/3) + (10/1) = (34/3)$, hence $C = 3/34 = 0.088$ µF.

(b) $1/C = (1/C_1) + (1/C_2) = (1/1) + (1/6) = 7/6$, hence $C = 6/7 = 0.86$ µ F. $C = C_1 + C_2 = 0.86 + 2 = 2.86$ µF. $1/C = (1/C_1) + (1/C_2) = (1/2.86) + (1/4) = (4 + 2.86)/(2.86 \times 4)$, hence $C = (2.86 \times 4)/6.86 = 1.67$ µF.

40 For entire circuit $Q = CV = 0.088 \times 50 = 4.4$ µC. This is charge on each capacitor. Thus $V = Q/C = 4.4/0.1 = 44$ V.

Chapter 3

1 A north pole in that like poles repel.

2 The left-hand end.

3 A magnet would experience a force. A current-carrying conductor would also experience a force.

4 $B = F/IL = 4 \times 10^{-3}/(2 \times 4 \times 10^{-2}) = 0.05$ T

5 (a) Upwards (b) Out of the paper (c) No force.

6 $F = BIL = 0.04 \times 5 \times 20 \times 10^{-2} = 0.04$ N

7 The vertical sides of the coil experiences forces which are in such directions as to cause the coil to be acted on by a torque. The directions of these forces could be like those shown in Figure 67. The horizontal sides of the coil do not experience any forces which could cause rotation. The coil does not keep on rotating because there is an opposing torque acting on the coil and that is provided by springs. The more the spring is coiled up by the movement of the coil the greater the resisting torque. When this resisting torque is equal to the torque resulting from the current in the coil then the coil stops rotating and stops at that particular angle. The angle is then related to the current in the coil and so the position of the pointer on the scale is a measure of the current.

8 When a current passes through the coil the sides of the coil parallel to the ends of the magnet experience forces which result in a torque, the other two sides of the coil do not experience forces which result in a torque. The coil can only keep rotating if the current through the side of the coil close to the north pole is always in the same direction, similarly the side of the coil close to the south pole must have the current constantly in the opposite direction. This is achieved by the use of the split ring commutator.

9 Torque = $NBILw\cos \theta = 200 \times 0.6 \times 0.100 \times 30 \times 10^{-3} \times 20 \times 10^{-3} \times \cos 45° = 5.09 \times 10^{-3}$ N m.

10 $\Phi = BA = 0.4 \times 6 \times 10^{-4} = 2.4 \times 10^{-4}$ Wb

11 $B = \Phi/A = 0.5 \times 10^{-3}/(10 \times 10^{-4}) = 0.5$ T

12 An e.m.f. is induced in a conductor as a result of the conductor cutting through magnetic flux.

13 Faraday's laws: an e.m.f. is induced in a circuit whenever the magnetic flux linked by that circuit changes. The magnitude of the induced e.m.f. is directly proportional to the rate of change of flux linked by the circuit. Lenz's law: the direction of an induced current is such that its effect would oppose the change in magnetic flux which gave rise to it.

14 You could describe an experiment like that illustrated in Figure 75. A very sensitive galvanometer would be required as it is unlikely that the current produced would be very large, or the induced e.m.f. very large.

15   $E = 0.5 \times 10^{-3}/0.1 = 5 \times 10^{-3}$ V

16   $E = 400 \times 6 \times 10^{-3}/0.3 = 8$ V

17   $E = 300 \times 2 \times 10^{-3}/0.1 = 6$ V

18   (a) A flux linked change and so an e.m.f. Current B to C
(b) A flux linked change and so an e.m.f. Current B to C
(c) A flux linked change and so an e.m.f. Current D to A
(d) A flux linked change and so an e.m.f. Current D to A

19   $E = BLv$, hence $v = E/BL = 5/(0.3 \times 0.40) = 41.7$ m/s

20   $E = BLv = 0.4 \times 0.4 \times 4 = 0.64$ V

21   (a) $E \propto L$, hence E is doubled.
(b) $E \propto B$, hence E is doubled.

22   See the text, the flux linked by the coil depends on its angle with respect to the constant magnetic field direction and thus the rotating coil is being linked by a continually varying flux. When the flux changes an e.m.f. is induced.

23   See Figure 83.
(a) When the coil is passing through the horizontal position
(b) When the coil is passing through the vertical position

Chapter 4

1   (a) m.m.f. can be defined as the product of the number of turns and the current for a coil.
(b) Reluctance is the m.m.f. across a circuit divided by the flux in that circuit.
(c) The relative permeability is the factor by which the permeability of free space must be multiplied to give the absolute permeability.

2   (a) Current (b) e.m.f. (c) Resistance

3   m.m.f. $= NI = 400 \times 300 \times 10^{-3} = 120$ A

4   Current $=$ m.m.f./$N = 200/200 = 1$ A

5   m.m.f. $= \Phi S$, hence $S = 300/(400 \times 10^{-6}) = 7.5 \times 10^5$ A/Wb

6   m.m.f. $= \Phi S$, hence $\Phi = 200/(2 \times 10^5) = 1 \times 10^{-3}$ Wb

7   m.m.f. $= \Phi S$

8   $\mu = \mu_r\mu_o$, hence $\mu_r = 1.2 \times 10^{-4}/(4\pi \times 10^{-7}) = 95.4$

9   $\mu = \mu_r\mu_o = 90 \times 4\pi \times 10^{-7} = 1.1 \times 10^{-4}$ T m A^{-1}

10   $S = L/(\mu_r\mu_o A) = 0.3/(100 \times 4\pi \times 10^{-7} \times 20 \times 10^{-4}) = 1.19 \times 10^6$ A/Wb

11   $S = L/(\mu_r\mu_o A) = 300 \times 10^{-3}/(80 \times 4\pi \times 10^{-7} \times 200 \times 10^{-6})$
$= 1.5 \times 10^{-7}$ A/Wb. But $\Phi =$ m.m.f./$S = 300/(1.5 \times 10^7) = 2.0 \times 10^{-5}$ Wb

12 $S = L/(\mu_r\mu_o A) = 400 \times 10^{-3}/(120 \times 4\pi \times 10^{-7} \times 30 \times 10^{-4})$
$= 8.8 \times 10^5$ A/Wb.
But m.m.f. $= \Phi S = 500 \times 10^{-6} \times 8.8 \times 10^5 = 440$ A.

13 (a) A (b) Wb (c) A/Wb (d) T m A^{-1} or H/m

14 Flux

15 (a) $3 \times 10^6 + 40 \times 10^6 = 43 \times 10^6$ A/Wb
(b) m.m.f. $= \Phi S$, hence $\Phi = 200/(43 \times 10^6) = 4.65 \times 10^{-6}$ Wb

16 Iron: $S = L/(\mu_r\mu_o A) = 200 \times 10^{-3}/(400 \times 4\pi \times 10^{-7} \times 6 \times 10^{-4})$
$= 6.63 \times 10^5$ A/Wb.
Air: $S = 15 \times 10^{-3}/(1 \times 4\pi \times 10^{-7} \times 6 \times 10^{-4})$
$= 1.99 \times 10^7$ A/Wb.
Total: $S = 6.63 \times 10^5 + 1.99 \times 10^7 = 2.06 \times 10^7$ A/Wb

17 Iron: $S = L/(\mu_r\mu_o A) = 300 \times 10^{-3}/(800 \times 4\pi \times 10^{-7} \times A)$
$= 298/A$
Air: $S = 30 \times 10^{-3}/(1 \times 4\pi \times 10^{-7} \times A) = 2.39 \times 10^4/A$
Total: $S = 298/A + 2.37 \times 10^4/A = 2.4 \times 10^4/A$.
As m.m.f. $= \Phi S$, then $\Phi = 300 \times 2/(2.4 \times 10^4/A) = 0.025/A$.
But $B = \Phi/A = 0.025$ T

18 (a) First part: $S = L/(\mu_r\mu_o A) = 220 \times 10^{-3}/(120 \times 4\pi \times 10^{-7} \times 4 \times 10^{-3}) = 3.65 \times 10^5$ A/Wb
Second part: $S = L/(\mu_r\mu_o A) = 100 \times 10^{-3}/(120 \times 4\pi \times 10^{-7} \times 2 \times 10^{-3}) = 3.32 \times 10^5$ A/Wb
Total: $S = 3.65 \times 10^5 + 3.32 \times 10^5 = 6.97 \times 10^5$ A/Wb
(b) $\Phi = $ m.m.f.$/S = 200/(6.97 \times 10^5) = 2.89 \times 10^{-4}$ Wb

19 There are two ways of dividing the figure, I am taking the wider section to be completely across the bottom. Thus mean length of narrower section $= 50 + 80 + 50 = 180$ mm. Mean length of wider section $= 90$ mm. The mean length is considered to be the length of the flux path and that is considered to be the centre line of the sections concerned. Thus for the narrower section: $S = L/(\mu_r\mu_o A) = 180 \times 10^{-3}/(500 \times 4\pi \times 10^{-7} \times 10 \times 10^{-3} \times 20 \times 10^{-3}) = 5 \times 10^6$ A/Wb
For the wider section: $S = 90 \times 10^{-3}/(500 \times 4\pi \times 10^{-7} \times 10 \times 10^{-3} \times 40 \times 10^{-3}) = 3.58 \times 10^5$ A/Wb
Total reluctance $= 5 \times 10^6 + 3.58 \times 10^5 = 53.58 \times 10^5$ A/Wb

20 Without the air gap: $S = L/(\mu_r\mu_o A) = \pi \times 350 \times 10^{-3}/(800 \times 4\pi \times 10^{-7} \times 300 \times 10^{-6}) = 3.65 \times 10^6$ A/Wb.
$\Phi = $ m.m.f.$/S = 800 \times 300 \times 10^{-3}/(3.65 \times 10^6) = 6.58 \times 10^{-5}$ Wb
With the gap: for the iron $S = (\pi \times 350 \times 10^{-3} - 0.5 \times 10^{-3})/(800 \times 4\pi \times 10^{-7} \times 300 \times 10^{-6}) = 3.64 \times 10^6$ A/Wb
For the air: $S = 0.5 \times 10^{-3}/(1 \times 4\pi \times 10^{-7} \times 300 \times 10^{-6}) = 1.33 \times 10^6$ A/Wb
Total $S = 3.64 \times 10^6 + 1.33 \times 10^6 = 4.97 \times 10^6$ A/Wb.
Hence $\Phi = $ m.m.f.$/S = 800 \times 300 \times 10^{-3}/(4.97 \times 10^6) = 4.83 \times 10^{-5}$ Wb

21 Magnetic leakage is when all the flux linked by the coil is not passed to other parts of the magnetic circuit. Magnetic fringing is when the flux in passing through, say, an air gap in a magnetic circuit, spreads out over a greater area.

22 The flux will be reduced.

23 (a) No change (b) Reduced

24 $H = NI/L = 400 \times 400 \times 10^{-3}/(200 \times 10^{-3}) = 800$ A/m

25 m.m.f. $= NI = 900 \times 500 \times 10^{-3} = 450$ A.
$H = $ m.m.f.$/L = 450/(\pi \times 100 \times 10^{-3}) = 1.43 \times 10^3$ A/m

26 $B = \mu_r\mu_oH = 150 \times 4\pi \times 10^{-7} \times 1000 = 0.19$ T

27 Ferromagnetic

28 Ferromagnetic

29 (a) 1.6 T
(b) $B/H = 1.6/2000 = 8 \times 10^{-4}$ T m A^{-1}
(c) $\mu = B/H = 8 \times 10^{-4}$ T m A^{-1}
(d) $\mu_r = \mu/\mu_o = 8 \times 10^{-4}/(4\pi \times 10^{-7}) = 637$

30 (a) 1.7 T
(b) $B/H = 1.7/5000 = 3.4 \times 10^{-4}$ T m A^{-1}
(c) $\mu = B/H = 3.4 \times 10^{-4}$ T m A^{-1}
(d) $\mu_r = \mu/\mu_o = 3.4 \times 10^{-4}/(4\pi \times 10^{-7}) = 271$

31 $H = $ m.m.f.$/L = NI/L = 1000 \times 600 \times 10^{-3}/(\pi \times 200 \times 10^{-3}) = $
955 A/m
$B = 1.3$ T

32 $H = 3000$ A/m. Hence m.m.f. $= NI = HL$
and $I = HL/N = 3000 \times \pi \times 120 \times 10^{-3}/600 = 1.9$ A

33 $B = \Phi/A = 0.6 \times 10^{-3}/(6 \times 10^{-4}) = 1$ T.
Hence for mild steel $H = 500$ A/m. For mild steel m.m.f. $= HL = $
$500 \times 0.25 = 125$ A
For air $H = B/\mu = 1/(4\pi \times 10^{-7}) = 2.5 \times 10^6$ A.
Hence for air m.m.f. $= HL = 2.5 \times 10^6 \times 0.5 \times 10^{-3} = 1250$ A.
Total m.m.f. $= 1375$ A

34 (a) 0.3 mWb, the same because they are in series
(b) For both metals $B = \Phi/A = 0.3 \times 10^{-3}/(4 \times 10^{-4}) = 0.75$ T
For the mild steel $H = 200$ A/m.
For mild steel m.m.f. $= HL = 200 \times 0.20 = 40$ A
For the cast iron $H = 5500$ A/m.
For the cast iron m.m.f. $= HL = 5500 \times 0.1 = 550$ A
Total m.m.f. $= 590$ A. Hence as m.m.f. $= NI$, the current
$I = 590/400 = 1.48$ A

35 (a) The relationship between the flux density and the magnetic field intensity during the magnetization of a material
(b) When the flux density becomes virtually constant and independent of the magnetic field intensity

36 (a) The flux density remaining when the magnetic field intensity has been reduced to zero

(b) The magnetic field intensity required to reduce the flux density to zero

37 It is narrower.

38 It encloses a smaller area.

39 The ferromagnetic material provides a lower reluctance path than through the air surrounding the component. Remember that reluctance is proportional to $1/\mu$ and so the bigger the permeability the smaller the reluctance.

Chapter 5

1 When a changing current in a conductor produces an induced e.m.f. in the same conductor

2 A circuit is said to have an inductance of one henry if when the current in the circuit changes at the rate of one ampere per second an e.m.f. of one volt is induced in it.

3 $E = L(dI/dt) = 500 \times 10^{-3} \times 20 = 6$ V

4 $L = E/(dI/dt) = 30/60 = 0.5$ H

5 $E = L(dI/dt) = 0.2 \times (8/0.05) = 32$ V

6 $E = L(dI/dt) = 0.5 \times (1.5/0.2) = 3.75$ V

7 $L = N(d\Phi/dI) = 800 \times 50 \times 10^{-6}/2 = 0.02$ H

8 $L = N\Phi/I = 2000 \times 2 \times 10^{-3}/6 = 0.67$ H

9 $L = N\Phi/I$, hence $\Phi = LI/N = 0.03 \times 1.5/500 = 9 \times 10^{-5}$ Wb

10 $L = N^2\mu\, A/l = 1000^2 \times 4\pi \times 10^{-7} \times 4 \times 10^{-4}/0.15 = 3.35 \times 10^{-3}$ H

11 $L \propto A/l \propto \frac{1}{4}/2 \propto 1/8$. The inductance decreases by a factor of 8.

12 The inductance of the circuit can result in a large induced e.m.f. when the current is switched off. This e.m.f. is proportional to the rate of change of current.

13 $E = L(dI/dt) = 0.2 \times 2/0.01 = 4$ V

14 The induced e.m.f. will be ten times greater and thus there is a greater chance of arcing across the switch contacts.

15 Energy $= \frac{1}{2}LI^2 = \frac{1}{2} \times 0.5 \times 3^2 = 2.25$ J

16 Energy $= \frac{1}{2}LI^2 = \frac{1}{2} \times 10 \times 2^2 = 20$ J

17 $L = N\Phi/I$, hence energy $= \frac{1}{2}LI^2 = \frac{1}{2} \times (N\Phi/I)I^2 = \frac{1}{2}N\Phi I$
$= \frac{1}{2} \times 400 \times 3 \times 10^{-3} \times 2 = 1.2$ J

18 When the current in one circuit changes, an e.m.f. is induced in another circuit. Such circuits are said to be coupled.

19 $E_1 = M(dI_2/dt) = 100 \times 10^{-3} \times 20 = 2$ V

20　$E_p/E_s = 6/60 = 1/10$. This is the turns ratio N_p/N_s.

21　$E_s/E_p = 10/1$, hence $E_s = 10 \times 240 = 2400$ V

22　$E_p/E_s = 5/1 = N_p/N_s$, hence $N_s = 400/5 = 80$

23　(a) $E_p/E_s = N_p/N_s = 600/50 = 12$, hence $E_s = 12/12 = 1$ V
　　(b) $I_s/I_p = N_p/N_s = 12$, hence $I_s = 12 \times 2 = 24$ A

Chapter 6

1　(a) The number of cycles per second
　　(b) The maximum value, either positive or negative, measured from the zero value
　　(c) One complete waveform before it repeats itself

2　(a) $2 \times 1.2 = 2.4$ A
　　(b) $f = 1/T$, hence $T = 1/f = 1/100 = 0.01$ s

3　(a) $\frac{1}{2} \times 10 = 5$ V (b) $f = 1/T = 1/0.05 = 20$ Hz

4　A direct current does not vary in its direction round the circuit, an alternating current does vary in direction.

5　(a) is a direct waveform (b) and (c) are alternating waveforms.

6　$2\pi \times 60/360 = \frac{1}{3}\pi$

7　(a) $v = V_m\sin\theta = 6\sin 45° = 4.24$ V
　　(b) $v = V_m\sin\theta = 20\sin 15° = 5.18$ V
　　(c) $\frac{1}{3}\pi = 60°$, hence $v = V_m\sin\theta = 12\sin 60° = 10.4$ V
　　(d) $0.4\pi = 360 \times 0.4\,\pi/2\pi = 72°$, hence $v = V_m\sin\theta = 6\sin 72° = 5.71$ V
　　(e) $v = V_m\sin 2\pi ft = 10\sin(2\pi \times 50 \times 0.003)$ $10\sin 0.3\,\pi = 10\sin 54° = 8.09$ V
　　(f) $v = V_m\sin(2\pi\,t/T) = 4\sin(2\pi\,0.005/0.03) = 4\sin 0.33\,\pi = 4\sin 59.4° = 3.44$ V

8　See Figure 197. The periodic time $= 1/f = 1/50 = 0.02$ s. This is the time for a complete cycle. The time for a half a cycle is thus 0.01 s and the time for a quarter of a cycle is 0.005 s.

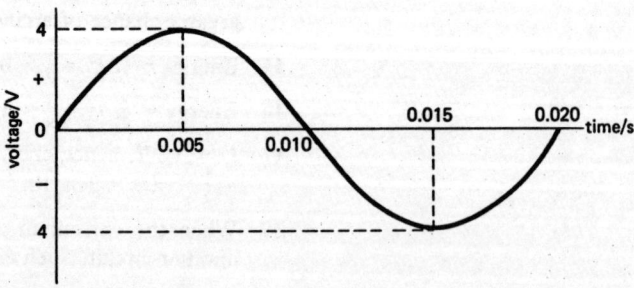

Figure 197

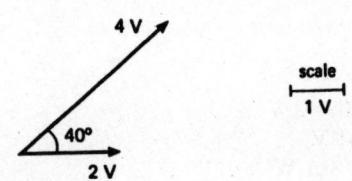

Figure 198

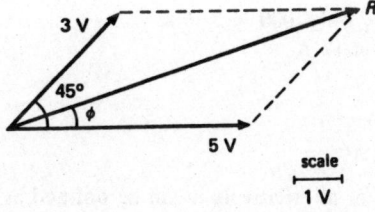

Figure 199

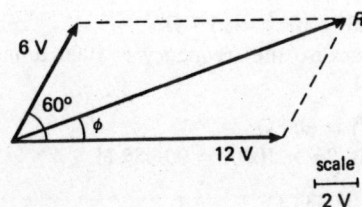

Figure 200

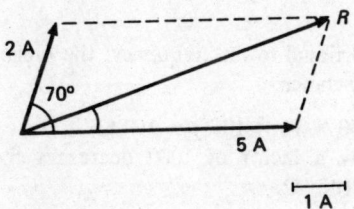

Figure 201

9 (a) Average = 30.5/6 = 5.1 A (b) Average = 18/6 = 3 V

10 Average = 0.637 × 2 = 1.274 A

11 (a) r.m.s. value = $\sqrt{(170.4/6)}$ = 5.33 A
(b) r.m.s. value = $\sqrt{(70/6)}$ = 3.42 V

12 r.m.s. value = 0.707 × 2 = 1.414 A

13 Using the mid-ordinate rule, or by just considering the obvious:
(a) The peak value (b) The peak value (c) One

14 Using the mid-ordinate rule and Figure 115(b):
(a) Root mean square value = 3.42 V
(b) Average value 3 V
(c) Form factor = 3.42/3 = 1.14

15 (a) Phasor 2 has the greatest peak value and phasor 1 leads by a phase angle of 30°.
(b) Phasor 1 has the greatest peak value and leads by a phase angle of 40°.
(c) Phasor 1 has the greatest peak value and leads by a phase angle of 90°.
(d) Phasor 2 has the greatest peak value, with phasor 1 leading by a phase angle of 180°.

16 See Figure 198.

17 See Figure 199. $R^2 = (3\sin 45°)^2 + (3\cos 45° + 5)^2$,
hence R = 7.43 V and tan φ = (3sin 45°)/(3cos 45° + 5) and φ = 16.6°.

18 $R^2 = 20^2 + 10^2$ and so R = 22.4 V. Tan φ = 20/10 and φ = 63.4°.

19 See Figure 200. $R^2 = (6\sin 60°)^2 + (6\cos 60° + 12)^2$,
hence R = 15.9 V. Tan φ = (6sin 60°)/(6cos 60° + 12) and φ = 19.1°.

20 See Figure 201. $R^2 = (2\sin 70°)^2 + (2\cos 70° + 5)^2$,
hence R = 5.99 A. Tan φ = (2sin 70°)/(2cos 70° + 5) and φ = 19.4°.

21 (a) $v = 2\sin 100\,\pi t$
(b) $v = 4\sin (100\,\pi t + \pi/2)$
(c) $v = 3\sin (100\,\pi t - \pi/4)$
(d) $i = 1.5\sin 100\,\pi t$

22 The output from a full wave rectifier uses both halves of the input whereas the output from a half wave rectifier uses only the positive half cycles.

23 The current only has a low resistance path through the rectifier in one direction, in the reverse direction there is very high resistance. Thus when an alternating voltage is applied, only the positive half cycles have a low resistance path and so give rise to a current. The negative half cycles have a very high resistance path and so give only a negligible current. Effectively the rectifier can be considered to be a switch which switches on for the positive half cycle and off for the negative half cycle.

Chapter 7

1 They are in phase.

2 (a) $V_m = I_m R = 100 \times 10^{-3} \times 20 = 2$ V
   (b) $V_{rms} = V_m/\sqrt{2} = 2/\sqrt{2} = 1.41$ V
   (c) Power $= V_{rms}^2/R = 1.41^2/20 = 0.1$ W

3 (a) $V_{rms} = V_m/\sqrt{2} = 24/\sqrt{2} = 17$ V
   (b) $I_{rms} = V_{rms}/R = 17/40 = 0.425$ A
   (c) Power $= I_{rms}^2 R = 0.425^2 \times 40 = 7.23$ W

4 (a) Power $= V_{rms}^2/R$, hence $R = 240^2/1000 = 57.6\ \Omega$
   (b) $I_{rms} = V_{rms}/R = 240/57.6 = 4.17$ A
   (c) $V_m = \sqrt{2}\ V_{rms} = 339.4$ V
   (d) $I_m = \sqrt{2}\ I_{rms} = 5.9$ A

5 The current lags the voltage by 90°.

6 Reactance is V_m/I_m or V_{rms}/I_{rms} or alternatively it can be defined as ωL.

7 X_L is directly proportional to the frequency, $X_L = 2\pi fL$.

8 (a) $X_L = 2\pi fL = 2\pi \times 50 \times 50 \times 10^{-3} = 15.7\ \Omega$
   (b) X_L is 1000 times greater because the frequency is 1000 times bigger, hence $X_L = 15.7$ kΩ

9 $X_L = V_{rms}/I_{rms} = 12/(20 \times 10^{-3}) = 600\ \Omega$.
   Hence $X_L = 2\pi fL$ and $L = 600/(2\pi \times 1000) = 0.0955$ H.

10 $X_L = V_{rms}/I_{rms} = 4/(12 \times 10^{-3}) = 333\ \Omega$

11 The current leads the voltage by 90°.

12 The reactance can be defined as $1/\omega C$ or $1/(2\pi fC)$ or in terms of V_m/I_m or V_{rms}/I_{rms}.

13 The reactance is inversely proportional to the frequency, the bigger the frequency the smaller the reactance.

14 (a) $X_C = 1/(2\pi fC) = 1/(2\pi \times 50 \times 16 \times 10^{-6}) = 199\ \Omega$
   (b) Increasing the frequency by a factor of 1000 decreases the reactance by 1000 to $199 \times 10^{-3}\ \Omega$.

15 $X_C = V_{rms}/I_{rms} = 20/(4 \times 10^{-3}) = 5000\ \Omega$. Hence as $X_C = 1/(2\pi fC)$ we have $C = 1/(2\pi fX_C) = 1/(2\pi \times 1000 \times 5000) = 3.18 \times 10^{-8}$ F

16 $X_C = V_{rms}/I_{rms} = 10/(0.5 \times 10^{-3}) = 2 \times 10^5\ \Omega$

17 $X_L = 2\pi fL = 2\pi \times 50 \times 0.2 = 62.8\ \Omega$.
   $Z = \sqrt{(X_L^2 + R^2)} = \sqrt{(62.8^2 + 50^2)} = 80.3\ \Omega$

18 $I_{rms} = V_{rms}/Z = 10/1000 = 0.01$ A

19 $X_L = 2\pi fL = 2\pi \times 50 \times 0.4 = 125.7\ \Omega$.
   $Z = \sqrt{(X_L^2 + R^2)} = \sqrt{(125.7^2 + 60^2)} = 139.3\ \Omega$.
   $I_{rms} = V_{rms}/Z = 240/139.3 = 1.72$ A.
   Tan $\phi = X_L/R = 125.7/60$ and so $\phi = 64.5°$.

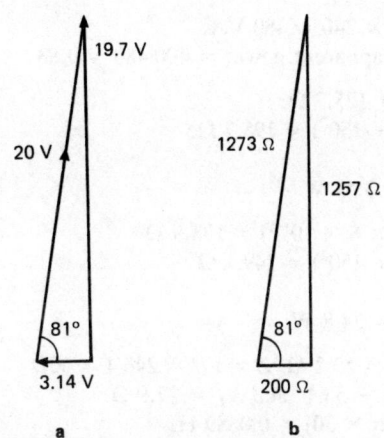

Figure 202 (a) *Voltage triangle*
(b) *Impedance triangle*

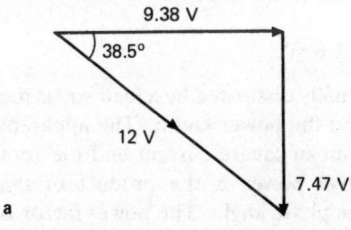

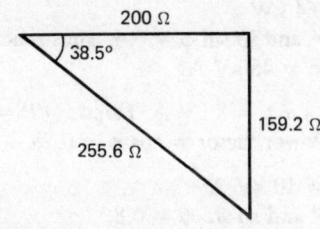

Figure 203 (a) *Voltage triangle*
(b) *Impedance triangle*

20 $Z = V_{rms}/I_{rms} = 150/3 = 50$ Ω. Increasing the frequency increases the reactance of the coil and so its impedance.

21 $R = V/I = 6/0.120 = 50$ Ω. $Z = V_{rms}/I_{rms} = 20/0.060 = 333.3$ Ω. $Z^2 = X_L^2 + R^2$, hence $X_L^2 = 333.3^2 - 50^2$ and so $X_L = 329.5$ Ω. $X_L = 2\pi fL$ hence $L = 329.5/(2\pi \times 100) = 0.52$ H.

22 $X_L = 2\pi fL = 2\pi \times 100 \times 2 = 1257$ Ω. $Z = \sqrt{(X_L^2 + R^2)} = \sqrt{(1257^2 + 200^2)} = 1273$ Ω.
$I_{rms} = V_{rms}/Z = 20/1273 = 0.0157$ A. Hence $V_R = IR = 0.157 \times 200 = 3.14$ V and $V_L = 0.0157 \times 1257 = 19.7$ V. Figure 202 shows the voltage and impedance triangles. Tan $\phi = X_L/R = 1257/200$ and so $\phi = 81.0°$.

23 $X_L = 2\pi fL = 2\pi \times 50 \times 0.3 = 94.2$ Ω. $Z = \sqrt{(X_L^2 + R^2)} = \sqrt{(94.2^2 + 50^2)} = 106.6$ Ω. $I_{rms} = V_{rms}/Z = 120/106.6 = 1.13$ A. For the inductor alone $Z = \sqrt{(X_L^2 + R^2)} = \sqrt{(94.2^2 + 20^2)} = 96.3$ Ω. Hence $V_L = IX_L = 1.13 \times 96.3 = 108.8$ V. $V_R = IR = 1.13 \times 30 = 33.9$ V

24 $X_C = 1/(2\pi fC) = 1/(2\pi \times 50 \times 2 \times 10^{-6}) = 1592$ Ω. $Z = \sqrt{(X_C^2 + R^2)} = \sqrt{(1592^2 + 100^2)} = 1595$ Ω

25 $Z = V_{rms}/I_{rms}$, hence $I_{rms} = 24/1200 = 0.02$ A

26 $X_C = 1/(2\pi fC) = 1/(2\pi \times 1000 \times 0.1 \times 10^{-6}) = 159.2$ Ω. $Z = \sqrt{(X_C^2 + R^2)} = \sqrt{(159.2^2 + 200^2)} = 255.6$ Ω. $I_{rms} = V_{rms}/Z = 12/255.6 = 0.0469$ A.
Tan $\phi = X_C/R = 159.2/200$ hence $\phi = 38.5°$.
This is the angle by which the voltage lags the current.

27 $V_R = IR = 0.0469 \times 200 = 9.38$ V. $V_C = IX_C = 0.0469 \times 159.2 = 7.47$ V. See Figure 203.

28 $X_C = 1/(2\pi fC) = 1/(2\pi \times 3000 \times 0.05 \times 10^{-6}) = 1061$ Ω. $Z = \sqrt{(X_C^2 + R^2)} = \sqrt{(1061^2 + 500^2)} = 1173$ Ω. $V_{rms} = ZI_{rms} = 1173 \times 15 \times 10^{-3} = 17.60$ V. Tan $\phi = X_C/R = 1061/500$ and so $\phi = 64.8°$.

29 $V_R = IR = 15 \times 10^{-3} \times 500 = 7.5$ V. $V_C = IX_C = 15 \times 10^{-3} \times 1061 = 15.9$ V. See Figure 204.

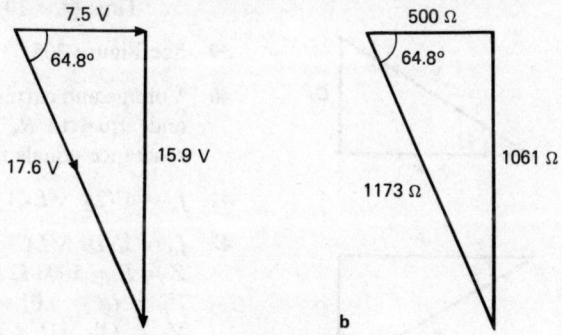

Figure 204 (a) *Voltage triangle* (b) *Impedance triangle*

30 (a) Apparent power = $IV = 2 \times 240 = 480$ VA.
(b) Power factor = true power/apparent power = $400/480 = 0.83$

31 $X_L = 2\pi fL = 2\pi \times 50 \times 0.4 = 125.7$ Ω.
$Z = \sqrt{(X_L^2 + R^2)} = \sqrt{(125.7^2 + 150^2)} = 195.7$ Ω.
$I = V/Z = 30/195.7 = 0.153$ A.
Power = $I^2R = 0.153^2 \times 150 = 3.51$ W

32 $X_C = 1/(2\pi fC) = 1/(2\pi \times 100 \times 8 \times 10^{-6}) = 198.9$ Ω.
$Z = \sqrt{(X_C^2 + R^2)} = \sqrt{(198.9^2 + 150^2)} = 249.1$ Ω.
$I = V/Z = 12/249.1 = 0.0482$ A.
Power = $I^2R = 0.0482^2 \times 150 = 34.8$ W

33 Power = I^2R, hence $R = 850/4^2 = 53.1$ Ω. $Z = V/I = 240/4 = 60$Ω.
$Z^2 = X_L^2 + R^2$, hence $X_L^2 = 60^2 - 53.1^2$ and $X_L = 27.9$ Ω.
$X_L = 2\pi fL$ and so $L = 27.9/(2\pi \times 50) = 0.0889$ H.

34 Power dissipated in capacitor is zero. $X_C = 1/(2\pi fC) = 1/(2\pi \times 50 \times 2 \times 10^{-6}) = 1592$ Ω.
$Z = \sqrt{(X_C^2 + R^2)} = \sqrt{(1592^2 + 800^2)} = 1782$ Ω.
$I = V/Z = 240/1782 = 0.135$ A.
Power = $I^2R = 0.135^2 \times 800 = 14.6$ W.

35 True power is the power that is actually dissipated by a load and is the product of the apparent power and the power factor. The apparent power is the product of the root mean square current and the root mean square voltage. The reactive power is the product of the apparent power and the sine of the phase angle. The power factor is the cosine of the phase angle, i.e. the ratio of the true power and the apparent power.

36 (a) $P = IV \cos\phi = 80 \times 0.8 = 64$ kW
(b) $\cos\phi = 0.8$, hence $\phi = 36.9°$ and so $\sin\phi = 0.6$. Thus reactive power = $IV \sin\phi = 80 \times 0.6 = 48$ kV Ar

37 $P = IV \cos\phi = 10$ kW, $Q = IV \sin\phi = 8$ kV A. Hence $Q/P = \tan\phi = 8/10$ and so $\phi = 36.7°$. Power factor = $\cos\phi = 0.78$

38 (a) $S = P/$power factor $= 6/0.6 = 10$ kV A.
(b) $\cos\phi = 0.6$ hence $\phi = 53.1°$ and so $\sin\phi = 0.8$.
Thus $Q = 10 \times 0.8 = 8$kV Ar

39 See Figure 205.

40 Voltage and current are in phase, the circuit impedance is a minimum and equal to R, the circuit current is a maximum, the inductive reactance equals the capacitive reactance.

41 $f_o = 1/(2\pi \sqrt{LC}) = 1/(2\pi \times \sqrt{0.05 \times 2 \times 10^{-6}}) = 503$ Hz

42 $f_o = 1/(2\pi \sqrt{LC}) = 1/(2\pi \times \sqrt{0.5 \times 16 \times 10^{-6}}) = 56.3$ Hz.
$Z = R = 1000$ Ω hence $I = V/Z = 10/1000 = 0.01$ A.
$V_R = IR = 0.01 \times 1000 = 10$ V = supply voltage.
$V_L = IX_L = I \times 2\pi fL = 0.01 \times 2\pi \times 56.3 \times 0.5 = 1.77$ V.
$V_C = IX_C = I/(2\pi fC) = 0.01/(2\pi \times 56.3 \times 16 \times 10^{-6}) = 1.77$ V.
Because it is resonance the V_L value must be the same as the V_C value.

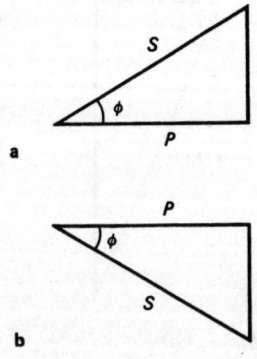

Figure 205

43 $f_0 = 1/(2\pi \sqrt{LC}) = 1/(2\pi \times \sqrt{0.2 \times 1 \times 10^{-6}}) = 356$ Hz.
$I = V/Z = V/R = 12/10 = 1.2$ A

44 $V_C = IX_C = I/(2\pi fC) = 1.2/(2\pi \times 356 \times 1 \times 10^{-6}) = 536$ V.
$V_L = V_C = 536$ V. $V_R = IR = 1.2 \times 10 = 12$ V.
Hence V across coil $= \sqrt{(V_L^2 + V_R^2)} = \sqrt{(536^2 + 12^2)} = 536$ V

45 $X_L = 2\pi fL = 2\pi \times 50 \times 0.5 = 157$ Ω.
$X_C = 1/(2\pi fC) = 1/(2\pi \times 50 \times 2 \times 10^{-6}) = 1592$ Ω.
$Z = \sqrt{((X_C - X_L)^2 + R^2)} = \sqrt{((1596 - 157)^2 + 100^2)} = 1442$ Ω.
$I = V/Z = 40/1442 = 0.0277$ A. $V_R = IR = 0.0277 \times 100 = 2.77$ V.
$V_C = IX_C = 0.0277 \times 1592 = 44.1$ V. $V_L = IX_L = 0.0277 \times 157 = 4.35$ V

46 $\omega_o = 1/\sqrt{(LC)} = 1/\sqrt{(2 \times 16 \times 10^{-6})} = 177$ rad/s.
$Q = 1/(\omega_o CR) = 1/(177 \times 16 \times 10^{-6} \times 100) = 3.53$.

47 $f_0 = 1/(2\pi \sqrt{LC}) = 1/(2\pi \sqrt{0.6 \times 16 \times 10^{-6}}) = 51.4$ Hz.
$I = V/Z = V/R = 50/10 = 5$ A

Chapter 8

1 Conductors 0.016 to 1 μΩ m; semiconductors 1 to 2000 Ω m; insulators 10^{13} to 10^{10} Ω m

2 $R = \rho L/A = 1 \times 10^{-6} \times 300 \times 10^{-3}/(\frac{1}{4}\pi \times 0.001^2) = 0.38$ Ω.

3 $L = 80 \times 2\pi \times 10 \times 10^{-3} = 5.03$ m.
$R = \rho L/A = 1 \times 10^{-6} \times 5.03/(\frac{1}{4}\pi \times 0.001^2) = 6.4$ Ω

4 A metal has the most free charge carriers and an insulator the least.

5 Because there is an increase in the number of free charge carriers

6 Because metal atoms have an electron which is distant enough from the positive nucleus to be very easily detached. In the solid at room temperature these electrons have become detached and so are available for conduction.

7 An insulator has all its electrons tightly held and large amounts of energy are needed to get them free.

8 In an intrinsic semiconductor there are as many free electrons as holes. In a doped semiconductor there are either more free electrons than holes or more holes than free electrons.

9 Doping results in conduction being primarily by either holes or electrons rather than equally by both.

10 The layer on either side of the junction which has lost holes or electrons because of the charge movement across the junction. The barrier potential is the potential that is produced across the junction by the movement of charge across it.

11 Forward bias is when a battery is connected so that it applies a potential difference across the junction which effectively cancels out the barrier potential. Reverse bias is when the applied potential difference reinforces the barrier.

12 With the positive side of the battery connected to the *p*-side of the junction, i.e. forward biased.

13 Only if the base is thin is it possible for the charge carrier movement through the forward biased junction to pass virtually entirely through the reverse biased junction.

14 (a) Forward biased (b) Reverse biased

15 See Figures 157 and 158.

16 See the text for a full description. Your answer should refer to a majority charge movement across the forward biased collector-base junction, the base being very thin and so virtually all this current continues through the reverse biased base-collector junction. Thus a current in the low resistance emitter-base circuit gives rise to a similar current in the high resistance base-collector circuit.

17 This is the current from the collector to the base due to the minority charge carriers.

18 Common emitter, common base, common collector.

19 Because the emitter is the common point in the input and output circuits.

Chapter 9

1 An analogue instrument gives a display where a movement, for instance, is related to the value of the quantity being measured. A digital instrument is where the display is a series of digits.

2 A deflecting torque is experienced by a current-carrying coil when in the magnetic field produced by a permanent magnet.

3 Without damping the pointer would not settle down to give a steady reading when a current is passed through the instrument but would oscillate about the current value. With suitable damping the pointer quickly settles down to give the steady reading. The damping is provided by the coil being wound on an aluminium frame. When this frame, effectively a loop of wire, moves in the magnetic field an e.m.f. is induced in it. This gives rise to a current, called an eddy current, in the frame. The direction of this current is such as to produce an effect opposing the effect which produced it.

4 A rectifier circuit can be used to rectify the current and give a unidirectional current for the instrument.

5 If a waveform with a different form factor is used the calibration is incorrect and so the readings indicated by the instrument will be in error.

6 (a) $R_g = I_g R_g/(I - I_g) = 10 \times 10^{-3} \times 15/(500 \times 10^{-3} - 10 \times 10^{-3}) = 0.306 \ \Omega$
$P = I_g^2 R_g + (I - I_g)^2 R_s = (10 \times 1^{-3})^2 \times 15 + (450 \times 10^{-3})^2 \times 0.306 = 0.0635 \ W$

(b) $R_g = I_g R_g/(I - I_g) = 10 \times 10^{-3} \times 15/(20 - 10 \times 10^{-3}) = 7.504 \times 10^{-3}\ \Omega$

$P = I_g^2 R_g + (I - I_g)^2 R_s = (10 \times 10^{-3})^2 \times 15 + (20 - 10 \times 10^{-3})^2 \times 7.504 \times 10^{-3} = 3\ \text{W}$

(c) $R_m = (V/I_g) - R_g = (0.5/10 \times 10^{-3}) - 15 = 35\ \Omega$.
Total resistance $= 50\ \Omega$ and so $P = I_g^2 R = (10 \times 10^{-3})^2 \times 50 = 5 \times 10^{-3}\ \text{W}$

(d) $R_m = (V/I_g) - R_g = (50/10 \times 10^{-3}) - 15 = 4985\ \Omega$.
Total resistance $= 5000\ \Omega$, hence $P = I_g^2 R = (10 \times 10^{-3})^2 \times 5000 = 0.5\ \text{W}$

7 Resistance

8 The instrument must be zeroed. This means adjusting a variable resistor so that the instrument gives a full scale current reading.

9 $E = I(R + R_v + R_g)$, hence $R_v = (E/I) - R - R_g = (1.5/100 \times 10^{-6}) - 1000 = 14\ 000\ \Omega$

(a) $R = (E/I) - R_v - R_g = (1.5/50 \times 10^{-6}) - 14\ 000 - 1000 = 15\ 000\ \Omega$

(b) $R = (1.5/25 \times 10^{-6}) - 14\ 000 - 1000 = 45\ 000\ \Omega$

10 Switch to d.c. or a.c.; select the highest range.

11 Reset the zero.

12 (a) The current passes through a coil and causes a magnetic field, which then causes two rods, or plates, to become magnetized and repel each other.

(b) The current passes through a coil and produces a magnetic field which attracts a plate towards it.

13 (a) Springs (b) An air dashpot

14 A non-linear scale and can be used with both a.c. and d.c.

15 Sensitivity = total resistance of voltmeter/voltage to give full scale deflection

16 The 'better' instrument is the one with a high sensitivity. Such an instrument has a high resistance for a particular voltage scale and so when connected across a resistor has the least effect on the current through the resistor and hence the potential difference across it.

17 Sensitivity = $10\ 000/20 = 500\ \Omega/\text{V}$

18 A voltmeter should have a higher resistance than an ammeter.

19 See Figure 176.

20 Systematic errors, observational errors, calibration errors

21 Any reading is only guaranteed to be accurate to within plus or minus 5 per cent of the full scale reading. Thus if the full scale reading was 100 V then the calibration of the instrument will only allow you to trust the reading to plus or minus 5 V.

22 The magnetic field is provided by a pair of current-carrying coils instead of a permanent magnet.

23 The magnetic field coils are supplied by the load current, the moving coil by a current related to the potential difference across the load. As the instrument gives a deflection proportional to the product of the currents through the field coil and the moving coil, the deflection is proportional to the power. Figure 188 shows how the instrument is connected to achieve the above.

24 (a) 40 V (b) 80 ms (c) $f = 1/80 \times 10^{-3} = 12.5$ Hz

25 $f_y = 2f_x = 200$ Hz

26 Look back through the text for a fuller description. Measure the distance along the X-axis for 1 cycle, convert to time using the time base calibration, then calculate frequency as being the reciprocal of this time.

27 A high resistance/impedance

28 0.001 V

29 $E_1/E_2 = L_1/L_2$, hence $E_1 = 1.205 \times 852/286 = 3.590$ V

30 $R_1/R_2 = L_1/L_2$, hence $R_1 = 6 \times 287/682 = 2.52 \ \Omega$

31 $IR/E_2 = L_1/L_2$, hence $I = 1.019 \times 658/(456 \times 5) = 0.294$ A

32 $P/Q = R/S$, hence $P = 50 \times 10.1/400 = 1.26 \ \Omega$

33 $P/Q = R/S$, hence $P = 100 \times 5.65 = 565 \ \Omega$

Index

a.c. circuits, 141
a.c. generator, 82
ampere-turn unit, 86
amplification, 184
amplitude, 121
analogue instruments, 186
apparent power, 161
arcing, 112
atoms, 175
average value, 126

barrier potential, 178
bias, 179
bipolar transistor, 180

calibration error, 199
capacitance, 48
capacitors in a.c. circuit, 148
capacitors in parallel, 59
capacitors in series, 59
capacitors:
    parallel plate, 51
    types of, 51
cathode ray oscilloscope, 201
charge, 40
circuit, magnetic, 85
circuit diagram, 11
coercivity, 105
coils in magnetic fields, 70
common base mode, 183
common collector mode, 183
common emitter mode, 183
conduction in solids, 174
conductors, 41, 173
configurations, transistor circuits, 183
core type transformer, 116
coulomb, unit, 41
current:
    concept, 9, 41
    direction of, 11
current balance, 68

damping, 187, 195
dashpot, 195
depletion layer, 178
diamagnetic, 98

dielectric strength, 55
digital instruments, 186
digital voltmeter, 205
diode, junction, 178
doping, 177
dynamometer instruments, 200

electric fields, 43
electric field strength defined, 43
electric force, 41
electric potential, 45
electromagnetic induction, 75
electron, 175
electronic voltmeter, 204
electromotive-force, 30, 47
e.m.f., 30, 47
energy stored:
    capacitor, 57
    inductor, 114
errors, instrument, 199

farad, unit, 50
Faraday's laws, 76
ferromagnetic, 98
field:
    electric, 43
    magnetic, 65
Fleming's rules:
    left-hand, 69
    right-hand, 77
flux, 74
flux density, 69
flux fringing, 95
flux leakage, 95
form factor, 132
free electrons, 174
fringing, flux, 95
full wave rectification, 140

galvanometer, 189
generator, a.c., 82
germanium, 189

half wave rectifier, 139
hard magnetic materials, 106
henry, unit, 88

holes, 176
hysteresis, 104

impedance, 151, 156
inductance, 108
inductance:
    mutual, 115
    self, 108
induction, electromagnetic, 75
inductor in an a.c. circuit, 145
inductor, energy stored in, 114
instrument errors, 199
insulators, 41, 173
internal resistance, 30
intrinsic semiconductors, 176

junction diode, 178

Kirchhoff's laws, 31

leakage, flux, 95
Lenz's law, 76
Lissajou's figures, 203

magnetic circuit, 85
magnetic field, 65
magnetic field intensity, 96
magnetic field strength, 96
magnetic flux, 74
magnetic flux density, 69, 74
magnetic force, 67
magnetic materials, 98
magnetic screening, 107
magnetization curve, 99
magnetizing force, 96
magnetomotive-force, 85
maximum value, 121
mean value, 127
mid-ordinate rule, 127
minority carriers, 179
m.m.f., 85
motor, d.c., 71
moving coil instrument, 71, 187
moving iron instrument, 195
multiplate capacitor, 54
multiplier, meter, 190
multiples of units, 16

multirange meter, 193
mutual inductance, 115

Neumann's equation, 76
n-type semiconductor, 177

observational error, 199
ohmmeter, 192
Ohm's law, 11

parallel circuits, 22
paramagnetic, 98
peak value, 121
permeability, 87
permittivity, 52
phase angle, 134
phasors, 133
potential, electric, 44
potential difference, concept, 9
potential divider, 21
potential gradient, 46
potentiometer, 21, 205
power, apparent, 161
power factor, 160
power in a.c. circuit, 160
power in d.c. circuit, 36
power triangle, 164
p-type semiconductor, 177

radian, 123
ranges of meter, 189
reactance:
    capacitive, 148
    inductive, 146
reactive power, 165
reactive voltamps, 165
rectification, 139
rectifier, 180
relative permittivity, 52
reluctance, 87
remanent flux density, 105
resistance:
    defined, 10
    internal, 30
resistivity, 172
resistor in a.c. circuit, 141
resistors in parallel, 23
resistors in series, 19
resonance, 167

screening, magnetic, 107
self inductance, 108
semiconductors, 173, 176
sensitivity, 197
series circuit, 19
series LCR circuit, 166

series magnetic circuit, 91
shell type transformer, 116
shunt, 189
silicon, 176
sine wave, 122
soft magnetic materials, 106
solenoid, 66
systematic errors, 199

tesla unit, 74
transformer, 116
transistor, bipolar, 180

units, multiples of, 16

variable capacitor, 54
voltage, concept, 10
voltmeter:
    digital, 205
    electronic, 204
voltmeter sensitivity, 197

watt, unit, 36
wattmeter, 200
waveforms, 120
weber, unit, 74
Wheatstone bridge, 208